The First Snap-Fit Handbook

Hanser Books Sponsored by SPE

Belofsky, Plastics: Product Design and Process Engineering
Bonenberger, The First Snap-Fit Handbook
Brostow/Corneliussen, Failure of Plastics
Chan, Polymer Surface Modification and Characterization
Charrier, Polymeric Materials and Processing – Plastics, Elastomers and Composites
Chung, Extrusion of Polymers
Del Vecchio, Understanding Design of Experiments: A Primer for Technologists
Ehrig, Plastics Recycling
Ezrin, Plastics Failure Guide
Gordon, Total Quality Process Control for Injection Molding
Gruenwald, Plastics: How Structure Determines Properties
Jones, Guide to Short Fiber Reinforced Plastics
Lee, Blow Molding Design Guide
Macosko, Fundamentals of Reaction Injection Molding
Malloy, Plastic Part Design for Injection Molding
Matsuoka, Relaxation Phenomena in Polymers
Menges/Mohren, How to Make Injection Molds
Michaeli, Extrusion Dies for Plastics and Rubber
Michaeli/Greif/Kaufmann/Vossebürger, Training in Plastics Technology
Michaeli/Greif/Kretzschmar/Kaufmann/Bertuleit, Training in Injection Molding
Neuman, Experimental Strategies for Polymer Scientists and Plastics Engineers
Osswald, Polymer Processing Fundamentals
Progelhof/Throne, Polymer Engineering Principles
Rauwendaal, Polymer Extrusion
Rees, Mold Engineering
Rosato, Designing with Reinforced Composites
Rotheiser, Joining of Plastics
Saechtling, International Plastics Handbook for the Technologist, Engineer and User
Stevenson, Innovation in Polymer Processing: Molding
Tucker, Fundamentals of Computer Modeling for Polymer Processing
Ulrich, Introduction to Industrial Polymers
Wright, Injection/Transfer Molding of Thermosetting Plastics
Wright, Molded Thermosets: A Handbook for Plastics Engineers, Molders and Designers

Paul R. Bonenberger

The First Snap-Fit Handbook

Creating Attachments for Plastic Parts

HANSER

Hanser Publishers, Munich
Hanser Gardner Publications, Inc., Cincinnati

The Author:
Paul R. Bonenberger, Rochester, MI 48307, USA

Distributed in the USA and in Canada by
Hanser Gardner Publications, Inc.
6915 Valley Avenue, Cincinnati, Ohio 45244-3029, USA
Fax: (513) 527-8950
Phone: (513) 527-8977 or 1-800-950-8977
Internet: http://www.hansergardner.com

Distributed in all other countries by
Carl Hanser Verlag
Postfach 86 04 20, 81631 München, Germany
Fax: + 49 (89) 98 12 64

The use of general descriptive names, trademarks, etc., in this publication, even if the former are not especially identified, is not to be taken as a sign that such names, as understood by the Trade Marks and Merchandise Marks Act, may accordingly be used freely by anyone.

While the advice and information in this book are believed to be true and accurate at the date of going to press, neither the authors nor the editors nor the publisher can accept any legal responsibility for any errors or omissions that may be made. The publisher makes no warranty, express or implied, with respect to the material contained herein.

Library of Congress Cataloging-in-Publication Data
Bonenberger, Paul R.
The first snap-fit handbook : creating attachments for plastic parts/Paul R. Bonenberger.
    p. cm.
Includes bibliographical references and index.
ISBN 1-56990-279-8 (hardcover)
    1. Assembly-line methods. 2. Fasteners—Design and construction. 3. Plastic—Molding. 4. Production engineering. I. Title.
TS178.4.B64 2000
621.8′8-dc21   00-035080

Die Deutsche Bibliothek - CIP-Einheitsaufnahme
Bonenberger, Paul R.:
The first snap fit handbook : creating attachments for plastic parts/
Paul R. Bonenberger. - Munich : Hanser, Cincinnati : Hanser/Gardner, 2000
    ISBN 3-446-21230-2

All rights reserved. No part of this book may be reproduced or transmitted in any form or by any means, electronic or mechanical, including photocopying or by any information storage and retrieval system, without permission in writing from the publisher.

© Carl Hanser Verlag, Munich 2000
Typeset in the UK by Techset Composition Ltd., Salisbury
Printed and bound in Germany by Druckhaus "Thomas Müntzer", Bad Langensalza

Foreword

Over the past decade we have seen a complete redefinition of the expected outcome of design for manufacture in the product development process. The term, design for manufacture (DFM), was often applied to a process of using rules or guidelines to assist in the design of individual parts for efficient processing. For this purpose the rule sets, or lists of guidelines, were often made available to designers through company specific design guides. Such information is clearly valuable to design teams who can make very costly decisions about the design of individual parts if these are made without regard to the capabilities and limitations of the required manufacturing processes. However, if DFM rules are used as the main principles to guide a new design in the direction of manufacturing efficiency, then the result will usually be very unsatisfactory. The end result of this guidance towards individual part simplicity will often be a product with an unnecessarily large number of individual functional parts, with a corresponding large number of interfaces between parts, and with a large number of associated items for connecting and securing. At the assembly level, as opposed to the manufactured part level, the resulting product will often be very far from optimal with respect to total cost or reliability.

The alternative approach to part-focused DFM, is to concentrate initially on the structure of the product and try to reach team consensus on the design structure which is likely to minimize cost when assembly as well as part manufacturing costs are considered. With this goal in mind, Design for Assembly (DFA) is now most often the first stage in the design for manufacture evaluation of a new product concept. The activity of DFA naturally guides the design team in the direction of part count reduction. DFA challenges the product development team to reduce the time and cost required for assembly of the product. Clearly, a powerful way to achieve this result is to reduce the number of parts which must be put together in the assembly process. DFA is a vehicle for questioning the relationship between the parts in a design and for attempting to simplify the structure through combinations of parts or features, through alternative choices of securing methods, or through spatial relationship changes.

An important role of DFA is to assist in the determination of the most efficient fastening methods, for the necessary interfaces between separate items in a design. This is an important consideration since separate fasteners are often the most labor-intensive group of items when considering mechanical assembly work. To reduce the assembly cost of dealing with separate fasteners, fastening methods, which are an integral part of functional items, should always be considered. For plastic molded parts, well-designed snap fits of various types can provide reliable high-quality fastening arrangements, which are extremely efficient for product assembly. It is not an overstatement to claim that snap-fitted assembly structures have revolutionized the manufacturing efficiency of almost all categories of consumer products.

In this context, *The First Snap-Fit Handbook* by Paul Bonenberger provides an extremely valuable resource for product development teams. The concept of complete snap-fit attachment systems, rather than isolated analyses of the mechanics of the snap-fit elements, represents a major advance in the design of integral plastic attachment methods. This concentration on "attachment level" rather than snap-fit "feature level" design has been developed and tested by Paul Bonenberger through years of solving attachment problems with product development teams at General Motors Corporation. This handbook contains the best blend of analysis and real-world design experience.

Wakefield, Rhode Island *Peter Dewhurst*

Preface

This book is a reference and design handbook for the attachment technology called snap-fits or sometimes, integral attachments. Its purpose is to help the reader apply snap-fit technology effectively to plastic applications. To do this, it arranges and explains snap-fit technology according to an Attachment Level^{TM} knowledge construct. The book is intended to be a user-friendly guide and practical reference for anyone involved with plastic part development and snap-fits. It is called "The First Snap-Fit Handbook" for two reasons: I believe it is the first book written that is devoted exclusively to snap-fits. I also hope it leads to increased interest and more books on the subject.

The reader should consider this book to be a "good start" in the ongoing process of understanding and organizing snap-fit technology. There is much more to be done, but one must begin somewhere. Although the original "attachment level" construct (created in 1990 and 1991) has proven to be fairly robust and complete, many details have evolved over the years as I learned more about the topic. The construct will continue to evolve and I encourage and welcome reader's comments on the subject; they will certainly help in the process.

My interest in the subject of snap-fits grew out of a very real need at General Motors. As a long-time fastening expert, I had typically been involved with threaded fasteners and traditional mechanical attachments. In the late 1980's and early 90's, as GM embraced design for manufacturing and assembly, the philosophies of Dr. Geoffrey Boothroyd and Dr. Peter Dewhurst [*Product Design for Manufacture and Assembly*, 1988, G. Boothroyd and P. Dewhurst, Department of Industrial and Manufacturing Engineering, University of Rhode Island, Kingston, RI] were formally adopted as the corporate direction and rolled out in a series of intensive training/workshop sessions. As a result, product designers and engineers began looking for alternatives to traditional loose fasteners, including threaded fasteners. Snap-Fit attachments immediately became popular but we soon discovered that there was little design information available in the subject. Calculations for cantilever hook performance could be found in many supplier design guides or as software but beyond that, no general snap-fit attachment expertise was captured in design or reference books. GM needed to bootstrap itself to a level of snap-fit expertise that was not written down anywhere. An intensive study of snap-fit applications resulted and eventually patterns of good design practices began to emerge. A "systems level" understanding of snap-fit attachments began to grow.

I called this systems level organization of snap-fits "attachment level" to emphasize its focus on the interface as a whole and to distinguish it from the traditional "feature level" approach. I have been teaching about snap-fits according to this attachment level model since 1991. The reaction after each class has been that attendees had indeed reached a new or better understanding of snap-fits. I trust and hope this book will have the same results for the reader.

The Attachment Level Construct (ALC) was only a personal vision in 1990. I believed it had potential and that it represented a unique approach to understanding snap-fit applications but I needed much more to make it reality. First was verification that I was not just reinventing or paraphrasing some existing but obscure snap-fit design practices; an extensive literature search verified that systems-level snap-fit practices were not documented anywhere. I also needed impartial validation that the model was indeed useful and worth pursuing. A colleague, Mr. Dennis Wiese who was Manager of the Advanced Product Engineering Body Components Group at that time, provided that initial validation. He also gave moral support and generously provided resources including his own engineers and significant amounts of his own time for debate and discussion of the fledgling snap-fit design methodology. Those discussions, sometimes "lively" and always useful, drove the insights that helped shape the original attachment level model. Dennis was certainly the "mid-wife" of the attachment level approach and I cannot thank him enough for his help. Other GM people involved with the infant methodology included Florian Dutke, Tom Froling, Daphne Joachim, Colette Kuhl, Chris Nelander, Tom Nistor, Tim Rossiter and Teresa Shirley.

Finally, Mike Carter, of GM University, deserves special thanks because in early 1990 he called me up and asked, "What are you fastening guys going to do about too many loose fasteners in our products?" That phone call was the beginning of my involvement with design for assembly. Mike, here is your answer.

As pressure of other work grew, the development team dwindled back to one (me). In 1992, Tony Luscher the project manager of a planned snap-fit program at Rensselaer Polytechnic Institute and I learned of each other's work and made contact (once again, thanks to Mike Carter). The RPI program was originally designed around feature level research but Tony enthusiastically embraced the concept of attachment level thinking. Tony, with the concurrence of Dr. Gary Gabrielle, the project leader, modified the RPI program to include some aspects of the attachment level method. Tony's technical insights, contributed during many hours of personal discussion and through exchange of correspondence, helped drive more refinements to the method. Under his guidance some work to apply and extend the methodology occurred under the RPI program. Tony is now a professor at the Ohio State University and he has carried his interest and enthusiasm for the subject to his new job. We continue to exchange ideas on the subject. Tony and I share a long-term vision for snap-fit technology: that attachment level thinking will lead to evolution of the snap-fit design and development process from an art to an engineering science.

The original motivation for the attachment level work was to provide support for Design for Manufacturing and Design for Assembly initiatives at General Motors. Joe Joseph, then the Director of the GM DFM Knowledge Center, supported my early efforts both verbally and by providing a site for snap-fit training classes. This also provided the kind of validation needed to justify continued efforts to develop the methodology. Joe is now Dean of the Engineering College of the GM University and he continues to provide valued moral support. The patience and support of Jim Rutledge, Dave Bubolz, and Roger Heimbuch is also greatly appreciated. They provided an environment in which ongoing development work could flourish and gave me much encouragement. Tony Wojcik, now with Delphi

Automotive Systems, also deserves thanks because he first sent a publisher my way. That marked the beginning of the snap-fit book project.

I must also acknowledge the creative people who designed and developed the numerous snap-fit applications I have studied. In products from around the world, the level of cleverness and creativity evident in many snap-fits is truly impressive. My admiration for and fascination with these designs helped to drive the original ideas behind the Attachment Level Construct in the following manner:

- *Observation*: There are many clever, well-designed and complex snap-fit applications in existence; there are also many poor snap-fits.
- *Hypothesis*: Many snap-fit designers must possess *tacit* knowledge that allows them to develop good snap-fits, others do not.
- *Problem*: Snap-Fit application design information could not be found as documented knowledge. Principles of good snap-fit application design were not written down anywhere.
- *Solution*: Discover the information and define it. Study successful snap-fit applications and look for patterns of good design practices. Capture and organize the concepts behind good snap-fit design.
- *Result*: A deep understanding of snap-fit concepts and principles organized in a knowledge construct.

I cannot claim credit for the vast majority of the clever snap-fit applications or concepts I describe here. Most were found on existing products or inspired by products. I simply interpreted them, inferred a logical process by which they could have been developed, and organized them into a knowledge structure. The only new "invention" here is the construct itself. Hopefully, it will inspire readers to create their own product inventions.

My wife and son have provided endless encouragement and understanding through the long process of writing this book, putting up with my long hours at the computer and tolerating (barely) my monopolization of same.

With thanks and appreciation to all.

Rochester, Michigan *Paul Bonenberger*

Contents

1 Snap-Fits and the Attachment Level™ Construct 1

   1.1 Introduction 1
   1.2 Reader Expectations 2
   1.3 Snap-Fit Technology 4
   1.4 Feature Level and Attachment Level 6
   1.5 Using this Book 8
       1.5.1 The Importance of Sample Parts 8
       1.5.2 Snap-Fit Novices 8
       1.5.3 Experienced Designers 10
       1.5.4 Design for Assembly Practitioners 10
   1.6 Chapter Synopses 10
   1.7 Extending the ALC to Other Attachments 11
   1.8 Summary 11
       1.8.1 Important Points in Chapter 1 12

2 Overview of the Attachment Level™ Construct 14

   2.1 Introduction 14
   2.2 The Key Requirements 14
       2.2.1 Strength 16
       2.2.2 Constraint 17
           2.2.2.1 Improper Constraint 19
       2.2.3 Compatibility 20
       2.2.4 Robustness 21
   2.3 Elements of a Snap-Fit 25
       2.3.1 Function 27
           2.3.1.1 Action 27
           2.3.1.2 Attachment Type 27
           2.3.1.3 Retention 28
           2.3.1.4 Lock Type 29
           2.3.1.5 Function Summary 29
       2.3.2 Basic Shapes 29
           2.3.2.1 Mating Part and Base Part 31
           2.3.2.2 Basic Shape Descriptions 31
           2.3.2.3 Basic Shape Summary 33
       2.3.3 Engage Direction 35

	2.3.4 Assembly Motion	38
	2.3.5 Constraint Features	40
	2.3.5.1 Locator Features	40
	2.3.5.2 Lock Features	42
	2.3.6 Enhancements	43
	2.3.7 Elements Summary	44
2.4	Summary	45
	2.4.1 Important Points in Chapter 2	46
	2.4.2 Important Design Rules Introduced in Chapter 2	46

3 Constraint Features 47

3.1 Introduction 47
3.2 Locator Features 47
    3.2.1 Locator Styles 48
        3.2.1.1 Lug 48
        3.2.1.2 Tab 48
        3.2.1.3 Wedge 49
        3.2.1.4 Cone 49
        3.2.1.5 Pin 50
        3.2.1.6 Catch 50
        3.2.1.7 Surface 50
        3.2.1.8 Land 50
        3.2.1.9 Edge 51
        3.2.1.10 Hole 52
        3.2.1.11 Slot 52
        3.2.1.12 Cutout 52
        3.2.1.13 Living Hinge 54
    3.2.2 Design Practices for Locator Pairs 55
        3.2.2.1 Terminology 55
        3.2.2.2 Locator Pairs, Constraint and Strength 57
        3.2.2.3 Locator Pairs and Ease of Assembly 61
        3.2.2.4 Locator Pairs and Dimensional Control 62
        3.2.2.5 Locator Pairs and Mechanical Advantage 64
        3.2.2.6 Locator Pairs and Compliance 65
        3.2.2.7 Locators Summary 67
3.3 Lock Features 67
    3.3.1 Lock Feature Styles 67
    3.3.2 Cantilever Beam Locks 68
        3.3.2.1 The Deflection Mechanism 69
        3.3.2.2 The Retention Mechanism 70
        3.3.2.3 Cantilever Lock Examples 74
        3.3.2.4 Locators as Cantilever Locks 74
        3.3.2.5 Lock Pairs 77
        3.3.2.6 Cantilever Lock Assembly Behavior 78

	3.3.2.7 Cantilever Lock Retention and Disassembly Behavior	80
3.3.3	Planar Locks	84
3.3.4	Trap Locks	85
	3.3.4.1 Trap Assembly Behavior	87
	3.3.4.2 Trap Retention and Disassembly	87
	3.3.4.3 Traps and Lock Efficiency	89
3.3.5	Torsional Lock	90
3.3.6	Annular Lock	91
3.3.7	Lock Pairs and Lock Function	91

3.4 Summary ... 92
   3.4.1 Important Points in Chapter 3 ... 92
   3.4.2 Design Rules Introduced in Chapter 3 ... 93

4 Enhancements ... 95

4.1 Introduction ... 95
4.2 Enhancements for Assembly ... 96
   4.2.1 Guidance Enhancements ... 96
      4.2.1.1 Guides ... 97
      4.2.1.2 Clearance ... 98
      4.2.1.3 Pilots ... 99
   4.2.2 Product Example 1 ... 100
   4.2.3 Product Example 2 ... 102
   4.2.4 Product Example 3 ... 102
   4.2.5 Operator Feedback ... 104
   4.2.6 Product Example 3 Revisited ... 108
   4.2.7 Assembly Enhancements Summary ... 108
4.3 Enhancements for Activating and Using Snap-Fits ... 109
   4.3.1 Visuals ... 109
   4.3.2 Assists ... 111
   4.3.3 User Feel ... 113
4.4 Enhancements for Snap-Fit Performance ... 114
   4.4.1 Guards ... 114
   4.5.2 Retainers ... 115
   4.4.3 Compliance ... 115
      4.4.3.1 Compliance through Local Yield ... 117
      4.4.3.2 Compliance through Elasticity ... 119
      4.4.3.3 Isolators ... 119
   4.4.4 Back-Up Locks ... 119
4.5 Enhancements for Snap-Fit Manufacturing ... 120
   4.5.1 Process-Friendly ... 121
   4.5.2 Fine-Tuning ... 125
4.6 Summary ... 128
   4.6.1 Important Points in Chapter 4 ... 128
   4.6.2 Design Rules Introduced in Chapter 4 ... 131

5 Other Snap-Fit Concepts 135

    5.1 The Importance of Constraint. 135
        5.1.1 Constraint Review 136
        5.1.2 Constraint Principles. 136
            5.1.2.1 Perfect Constraint 136
            5.1.2.2 Proper Constraint 139
            5.1.2.3 Proper Constraint in Less than 12 DOM 139
            5.1.2.4 Under-Constraint 139
            5.1.2.5 Over-Constraint 140
            5.1.2.6 General Constraint Rules 141
        5.1.3 The Constraint Worksheet 143
        5.1.4 Additional Comments on Constraint 151
    5.2 Lock Decoupling 151
        5.2.1 The Lock Feature Paradox. 151
        5.2.2 Decoupling Examples. 151
        5.2.3 Levels of Decoupling 153
            5.2.3.1 No Decoupling (Level 0) 153
            5.2.3.2 Level 1 Decoupling. 154
            5.2.3.3 Level 2 Decoupling. 155
            5.2.3.4 Level 3 Decoupling. 156
            5.2.3.5 Level 4 Decoupling. 156
        5.2.4 Decoupling Summary 159
    5.3 Summary 160
        5.3.1 Important Points in Chapter 5 160
        5.3.2 Design Rules Introduced in Chapter 5 160

6 Feature Design and Analysis 162

    6.1 Pre-Conditions for Feature Analysis 163
    6.2 Material Property Data Needed for Analysis 163
        6.2.1 Sources of Materials Data 163
        6.2.2 Assumptions for Analysis 164
        6.2.3 The Stress-Strain Curve. 165
        6.2.4 Establishing a Design Point 168
        6.2.5 Coefficient of Friction (μ) 172
        6.2.6 Other Effects 175
    6.3 Cantilever Hook Design Rules of Thumb 178
        6.3.1 Beam Thickness at the Base. 178
        6.3.2 Beam Length. 178
        6.3.3 Insertion Face Angle. 181
        6.3.4 Retention Face Depth 181
        6.3.5 Retention Face Angle 182
        6.3.6 The Threshold Angle 183
        6.3.7 Beam Thickness at the Retention Feature 183

6.3.8 Beam Width	184
6.3.9 Other Features	185
6.4 Initial Strain Evaluation	185
6.5 Adjustments to Calculations	187
6.5.1 Adjustment for Stress Concentration	188
6.5.2 Adjustment for Wall Deflection	189
6.5.3 Adjustment for Mating Feature Deflection	191
6.5.4 Adjustment for Effective Angle	194
6.5.4.1 Effective Angle for the Insertion Face	195
6.5.4.2 Effective Angle for the Retention Face	196
6.5.5 Adjustments Summary	197
6.6 Assumptions for Analysis	198
6.7 Using Finite Element Analysis	198
6.8 Determine the Conditions for Analysis	199
6.9 Cantilever Hook Analysis for a Constant Rectangular Section Beam	199
6.9.1 Section Properties and the Relation between Stress and Strain	199
6.9.2 Evaluating Maximum Strain	201
6.9.2.1 Adjusting Maximum Allowable Strain for Stress Concentrations	202
6.9.2.2 Calculating the Maximum Applied Strain in a Constant Section Beam	203
6.9.2.3 Adjusting the Calculated Strain for Deflection Magnification	203
6.9.3 Calculating Deflection Force	204
6.9.4 Adjusting for Mating Part/Feature Deflection	205
6.9.6 Determine Maximum Assembly Force	206
6.9.6.1 Determine the Effective Insertion Face Angle	206
6.9.7 Determine Release Behavior	207
6.9.7.1 Additional Retention Considerations	208
6.10 Cantilever Hook Tapered in Thickness	209
6.11 Cantilever Hook Tapered in Width	211
6.12 Cantilever Hook Tapered in Thickness and Width	212
6.13 Modifications to the Insertion Face Profile	212
6.14 Modifications to the Retention Face Profile	215
6.15 Other Feature Calculations	215
6.16 Summary	215
6.16.1 Important Points in Chapter 6	216
7 The Snap-Fit Development Process	**218**
7.1 Introduction	218
7.1.1 Concept Development vs. Detailed Design	219
7.1.2 A General Development Process	219
7.2 The Snap-Fit Development Process	224
7.2.1 Is the Application Appropriate for a Snap-Fit? (Step 0)	224

	7.2.2 Define the Application (Step 1)	228
	7.2.3 Benchmark (Step 2)	230
	7.2.3.1 Rules for Benchmarking	231
	7.2.4 Generate Multiple Attachment Concepts (Step 3)	232
	7.2.4.1 Select Allowable Engage Directions (Step 3.1)	233
	7.2.4.2 Identify All Possible Assembly Motions (Step 3.2)	234
	7.2.4.3 Engage Directions, Assembly Motions and Worker Ergonomics	236
	7.2.4.4 Select and Arrange Constraint Pairs (Step 3.3)	237
	7.2.4.5 Add Some Enhancement Features (Step 3.4)	242
	7.2.4.6 Select the Best Concept for Feature Analysis and Detailed Design (Step 3.5)	242
	7.2.5 Feature Analysis and Design (Step 4)	243
	7.2.5.1 Lock Alternatives	245
	7.2.6 Confirm the Design with Parts (Step 5)	249
	7.2.7 Fine-Tune the Design (Step 6)	250
	7.2.8 Snap-Fit Application Completed (Step 7)	250
7.3	Summary	250
	7.3.1 Important Points in Chapter 7	250

8 Diagnosing Snap-Fit Problems 254

8.1	Introduction	254
	8.1.1 Rules for Diagnosing Snap-Fit Problems	255
	8.1.2 Mistakes in the Development Process	255
8.2	Attachment Level Diagnosis	256
	8.2.1 Most Likely Causes of Difficult Assembly	257
	8.2.2 Most Likely Causes of Distorted Parts	257
	8.2.3 Most Likely Causes of Feature Damage	257
	8.2.4 Most Likely Causes of Loose Parts	258
8.3	Feature Level Diagnosis	258
8.4	Summary	263
	8.4.1 Important Points in Chapter 8	263

Index 265

1 Snap-Fits and the Attachment Level™ Construct

Any scientific discipline has a need for a specific language for describing and summarizing the observations in that area. [1]

1.1 Introduction

The traditional snap-fit design process has consisted of calculations for predicting the behavior of individual locking features; we can describe this as the *feature level* of snap-fit technology. For example, the cantilever hook feature (Fig. 1.1) has been a particularly popular subject of feature level research. To many product designers, the cantilever hook (a *feature*) represents the sum total of snap-fit technology. As necessary and important as it is, however, feature level knowledge alone can not address many of the problems faced by those who must develop snap-fit applications.

Particularly for the first-time snap-fit designer, feature calculations alone are not enough and they find themselves learning the subject through trial-and-error, an expensive and time-consuming proposition. An oft-repeated phrase is "snap together—snap apart". Unfortunately, one or two bad experiences with snap-fits may cause a designer or an organization to swear off (after *swearing at*) snap-fits forever. To remain competitive, companies must utilize all possible design strategies. To ignore snap-fits as a valid attachment strategy is a mistake.

Part-to-part fastening occurs across a joint or interface. To wait until a component design is completed and then to begin designing the attachments for that interface is to invite problems. Much of this book will focus on getting that initial interface concept right. Some studies [2, 3] have shown that much of the cost of a product (~70%) is determined during the concept development stages, not during the actual product design. Why should snap-fits be any different?

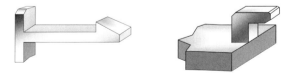

Figure 1.1 The cantilever hook is a common locking feature and the lug is a common locating feature

A comment sometimes heard about the attachment level approach is that it is "too basic". The response is always, "Yes it is basic, but that doesn't mean it isn't important. "In fact, because it is basic, it *must* be understood. Just because it is basic does not mean that it is widely understood and applied. Some snap-fits the author has seen can only be described as design disasters. Some never make it into production because they are so bad, representing wasted design time and lost opportunities for savings. Many others could be improved by applying these basic principles.

Dr. W. Edwards Deming [4] said, "Experience without theory teaches...nothing." The theory and fundamental knowledge provided by the Attachment Level™ Construct (ALC) can greatly improve the learning and understanding of snap-fit technology.

The Attachment Level™ Construct is simply a way of explaining the broad and varied world of snap-fits. It is a tool for organizing and capturing information and concepts. "...we create constructs by combining concepts and less complex constructs into purposeful patterns.... Constructs are useful for interpreting empirical data and building theory. They are used to account for observed regularities and relationships. Constructs are created in order to summarize observations and to provide explanations." [1]

A systematic way of thinking about attachments should appeal to designers, engineers, design-for-assembly practitioners, and technical trainers. Anyone wanting to develop improved mechanical attachments will benefit from attachment level thinking. With its help, the reader can more quickly reach an understanding of snap-fits that previously took years to acquire. Readers will also find that many of the ideas presented here can, and should, be applied to all mechanical attachments and interface designs, not just snap-fits. A discussion of how attachment level thinking can be extended to other attachments is included in Chapter 2.

Let us start with a common and traditional definition of a snap-fit. Shortly, we will refine it and present an attachment level definition of a snap-fit, but this will do for now.

> A snap-fit is a "built-in" or integral latching mechanism for attaching one part to another. They are commonly associated with plastic parts. A snap-fit is different from loose or chemical attachment methods in that it requires no additional pieces, materials or tools to carry out the attaching function.

In this chapter, we will introduce a systems approach to snap-fit technology and describe the fundamental differences between it and the traditional feature level way of thinking about snap-fits. Organization of the book is described and suggestions are made for its use.

This chapter also describes some differences between snap-fits and threaded fasteners. We should note here that the issue is not one of snap-fit technology versus threaded fastener technology. Neither is inherently good nor bad. Both have their place in product design based on informed selection and application of the best method for the design situation.

1.2 Reader Expectations

This book considers the snap-fit as an attachment *system*, (Fig. 1.2). This approach is based on an Attachment Level™ Construct (ALC), so named to emphasize its difference from traditional snap-fit design methods, and it is new.

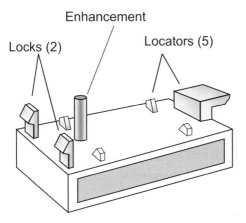

Figure 1.2 A snap-fit is a system of features interacting across a part-to-part interface

Because it is about snap-fits, many people hearing about the ALC for the first time assume that it is a variation of the feature level (i.e. calculation) approach to snap-fits. It is not. The reader should understand that this book is not primarily about mathematical analysis of feature behavior and is not a repetition of previously published feature level snap-fit information. While the book contains some feature level calculations, they are accorded relatively little space here because many references on that subject are already available. Likewise, plastic material properties and processing are dealt with only to the extent necessary to support attachment level understanding. Many excellent books and references are already available on those topics.

In short, this book is not about anything the reader is likely to expect in a book about snap-fits. It is also not a "cookbook" for snap-fits. The ALC leads to a rule-based snap-fit attachment development method and this book is primarily about learning and using those rules. The reader should expect to acquire a deep intuitive or gut-level understanding of snap-fits from reading this book and applying the principles of the ALC. Most importantly, the reader will learn how to *think* about snap-fits.

The book and the Attachment Level™ Construct are also not a collection of new snap-fit inventions. You will not find any new or revolutionary designs for snap-fit hardware. The only new "invention" here is the construct itself. The ideas in this book should, however, help the reader create their own snap-fit applications.

Five capabilities are necessary for an individual (or an organization) wanting to do a good job on snap-fits. They are *technical understanding*, *communication*, *attention to detail*, *spatial reasoning*, and *creativity*. As shown in Fig. 1.3, they tend to build upon each other. The ALC supports these capabilities in the following ways:

- Communication—The ALC provides a common and rational vocabulary for exchanging ideas and information about snap-fits. Any technical discipline requires a "language" if it is to be understood and used effectively.
- Technical understanding—The ALC organizes existing knowledge about snap-fits for easy understanding and use. It also supports capture and transfer of useful snap-fit

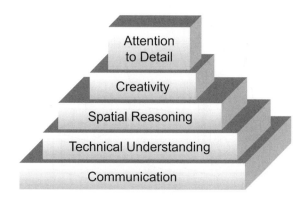

Figure 1.3 Snap-fit development requires five skills

knowledge and lessons-learned from one application to another. The organizing structure of the ALC also helps the user to grow in knowledge and add to their own technical understanding of snap-fits. Technical understanding also includes analytical capability for evaluating feature performance, the traditional feature level of snap-fit technology.
- Spatial reasoning—Snap-fit development is enhanced when the designer can visualize the interactions and behaviors of the parts to be joined as well as the features of the parts. The ALC provides a logical set of generic shapes and motions to enable this visualization.
- Creativity—The snap-fit development process (described in detail in Chapter 7) encourages creativity by supporting the generation of multiple attachment concepts for consideration by the designer.
- Attention to detail—The many details of snap-fit design are captured in a logical structure for the designer's consideration.

1.3   Snap-Fit Technology

Throughout this book, we will use the shorter term *snap-fit* rather than the term *integral attachment*.

The important criterion for a snap-fit is flexibility in the integral locking feature. As we will see, lock flexibility may be great or very small, depending on the lock style. Snap-fits are not limited to plastic parts. Effective snap-fits are also possible in metal-to-metal and plastic-to-metal applications. Keep this in mind as you read this book and look for opportunities to use snap-fits. Simply substitute the appropriate material properties and analytical procedures for the metal component(s) and features then proceed merrily on your way.

1.3 Snap-Fit Technology

Although commonly associated with parts made from plastic materials, snap-fits have been in existence long before plastics. Metal-to-metal snap-fits were and are popular—"snaps" on clothing, for example. Many styles of metal spring clips are essentially self-contained snap-fits.

Plastics, however, have made the snap-fit more practical and much more popular because of the relative flexibility of the material. Plastic processing technologies like injection molding have made production of complex shapes economically feasible. The advantages of ease of assembly and disassembly and the ever-increasing engineering capabilities of plastic materials now make the snap-fit a serious candidate for applications once considered the domain of threaded or other fasteners. Note that, while toys and small appliances have long made extensive use of snap-fits, the technology is now being applied extensively in the automotive components and electronics fields and is even being extended to structural applications. [5, 6, 7]

It is also very important to realize that experience with threaded fasteners, the most common method of mechanical attachment, is not transferable to designing snap-fit interfaces. New ways of thinking about functional requirements, component interfaces and attachments must be learned. That is so important, it bears repeating: *Experience with the most common method of mechanical attachment (threaded fasteners) does not transfer to designing snap-fit interfaces.*

Without intending any insult to threaded fastener technology, we can think of a threaded attachment as a "brute force" approach to connecting parts. The strength of the fastener can make it easy to ignore or forget many of the finer points of interface design and behavior. A retention problem can often be fixed by simply using a higher strength material for the fastener, tightening to a higher clamp load, specifying a larger fastener or adding more fasteners. Indeed, one of the major advantages of the loose fastener is that its strength is *independent* of the joined components. This is not the case with snap-fits.

With a snap-fit application, we do not have the luxury of selecting a fastener style, material and strength independent of the joined components. We must work with the material that has been selected for the parent components. Sometimes, attachment performance is a consideration in material selection but much of the time material selection is driven by other application considerations, not by the attachment requirements. The requirements and realities of part processing also restrict us, since the attachment features must be formed along with the part.

To make a snap-fit work, the subtleties of interface design and behavior must be understood and reflected in the design. In this sense a snap-fit application, of necessity, must be a more "elegant" method of attachment than a bolted joint.

Another point to keep in mind is that many, if not most, snap-fit designers are not materials experts. Anyone developing snap-fit applications should maintain very close contact with a polymer expert, preferably as early in the design process as possible. Maintaining a good relationship with a processing expert is also a good idea. Thus we see that coffee and donuts can be very useful tools in the snap-fit design process! The processing experts, in particular will appreciate your interest because they will ultimately have to produce your design.

The text refers frequently to *designers* of snap-fits. This does not refer to a job classification. The term means anyone who makes design decisions about snap-fits.

The term *snap-fit development* also includes much more than just analysis and detailed design of the snap-fit interface and components. It refers to all the steps in the process, from creating the initial concept through detailed analysis, design and testing.

1.4 Feature Level and Attachment Level

The designer must have a deep understanding of both the attachment and feature levels of snap-fit technology to ensure a good (reliable, easy-to-assemble and cost-effective) attachment. The attachment level is the more basic of the two because it provides for a fundamentally sound attachment *concept*, (Fig. 1.4). Once a good concept is established, feature level analysis is used to determine individual feature performance. If a good attachment concept is not established first, then even well designed features may fail. Furthermore, with respect to problem diagnosis, if the attachment or systems level causes of a problem are not understood, any attempt to fix that problem at the feature level will certainly be more expensive than necessary and possibly doomed to failure. The snap-fit designer should also have, at the least, a basic awareness of polymers and processing. If the designer is not an expert in these areas, finding someone with this expertise to provide input to the design is critical to success.

The name Attachment Level™ Construct is all-inclusive, referring to the entire approach to snap-fit development. It includes logical organization of snap-fit knowledge, attachment level and feature level terminology, design rules and the process for applying them. The feature level aspects of snap-fit design are integrated in the ALC and related areas such as plastic processing are captured where appropriate. Here are two attachment level definitions of a snap-fit. First the long definition:

> A snap-fit is a mechanical joining system where part-to-part attachment is accomplished with locating and locking features (constraint features) that are homogenous with one or the other of the components being joined. Joining requires the (flexible) locking features to move aside for

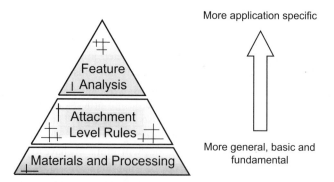

Figure 1.4 Snap-fit knowledge hierarchy

engagement with the mating part, followed by return of the locking feature toward its original position to accomplish the interference required to latch the components together. Locator features, the second type of constraint feature, are inflexible, providing strength and stability in the attachment. Enhancements complete the snap-fit system, adding robustness and user-friendliness to the attachment.

The shorter definition:

A snap-fit is an arrangement of compatible locators, locks and enhancements acting to form a mechanical attachment between parts, (Fig. 1.2).

We can see that thinking of a snap-fit as a system rather than as a feature moves us much closer to the realities of product applications, (Fig. 1.5). Thus attachment level design rules and guidelines have much more relevance to actual design problems and situations than do the feature level design rules alone.

We know an attachment level approach is fundamental to good snap-fit design because when attachment level requirements are not met, even an application with well-designed features is likely to have problems. In fact, as we study the causes of snap-fit failures and other problems we find that, in many cases, feature design is not the root cause of the problem. Feature failure may be a symptom of a more fundamental problem *which can only be solved at the attachment level*. Common plastic part problems likely to have attachment level causes include:

- Difficult assembly
- Feature damage or failure
- Part squeak and rattle
- Part warpage
- Loose parts

Of course, most of these problems may also result from poor feature design, but experience indicates that attachment level mistakes are the root cause or a contributing cause to most part attaching problems. Even feature damage and failure is often only a symptom of an attachment level problem.

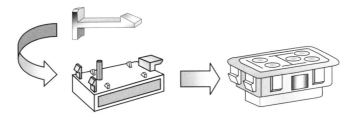

Figure 1.5 The attachment level is closer to the final product

1.5 Using this Book

Because this is the first book on this subject, it will prove to be far from perfect. I have tried to organize the chapters and sections so they flow in a reasonably logical manner. While some readers will find my choice of organization acceptable, I don't doubt that others will find it irritating. Some areas are treated in more detail than others. But, one has to start somewhere. I hope that the ideas presented here will lead to generation of additional ideas and constructive feedback that can be incorporated into continuous growth and improvement of snap-fit knowledge and of this book. The book can be read and used in many ways depending on need and interest. Reading Chapters 2 and 5 will give the casual reader a basic understanding of how snap-fits work. Figure 1.6 shows the overall layout of the book.

1.5.1 The Importance of Sample Parts

For everyone: Snap-fits are a highly spatial and visual topic. The best way, by far, to understand them is to hold them in your hands. It is highly recommended that the reader have some snap-fit applications available to study for reinforcement of the principles and ideas in the book. As you read, identify and classify the various features on these parts. Try to recognize the principles and rules that are being applied in the design. Snap-fit applications are everywhere; find them in toys, electronics, small appliances, vacuum cleaners, etc. They can be found in products as diverse as patio lamps, chemical sprayers, slot-car tracks and toilet tank shut-off valves. An excellent source of snap-fits is the Polaroid One-Step© camera that has been around in various styles for many years. Buy one new or pick one up at a garage sale and take it apart. They are 100% snap-fit and the variety and cleverness of the attachments is impressive. The IBM Pro-Printer© is also an excellent source of clever attachments.

One word of caution: The ideas and examples shown in the book were collected over many years from a wide variety of consumer products and applications. Examples are provided here as idea starters and illustrations of various principles. They are presented without consideration of specific patents on the product as a whole. In most cases, the original product identification has been lost. Individual commonly used features like cantilever hooks are not patented. However, an entire interface system that uses cantilever hooks, would be included in a patented design. Use the information in this book to create your own unique and patentable products.

1.5.2 Snap-Fit Novices

A novice in snap-fit design should read the book in chapter order. It is laid out in a logical manner for maximum understanding. Once familiar with the entire subject, solidify understanding by stepping through the development process (Chapter 7) with a few sample

1	Snap-fits and the Attachment Level Construct - p. 1
	Background to snap-fit technology and to the book itself.

2	Overview of the Attachment Level Construct - p. 14
	Explains the model used to organize the information presented in this book

3	Constraint and Constraint features - p. 47
	Describes the two fundamental features of a snap-fit attachment system.
	Locators - p. 47 Locks - p. 67

4	Enhancements - p. 95
	Describes additional desirable features of a snap-fit attachment system.
	Assembly p. 96 Activation p. 109 Performance p. 114 Manufacturing p. 120

5	Fundamental Concepts - p. 135
	Describes the two fundamental features of a snap-fit attachment system.
	Constraint - p. 135 Decoupling - p. 151

6	Snap-fit Feature Analysis - p. 162
	Basics of cantilever hook design and performance analysis

7	The Snap-fit Development Process - p. 218
	A logical step-by-step method for creating sound snap-fit designs.

8	Diagnosing Common Snap-fit Problems - p. 254
	Understanding the root cause of a problem before fixing the wrong thing.

Figure 1.6 Layout of the book

applications. A team approach to learning about snap-fits can be extremely effective; several people can study parts and, using attachment level terminology, discuss their good and bad points and behavior. This will encourage attachment level thinking and reinforce understanding of the terminology.

1.5.3 Experienced Designers

More experienced designers interested in details of product design will learn a practical development process (in Chapter 7) that will allow them to reach a better attachment design faster. Chapter 5 contains explanations of several deeper snap-fit concepts. Many experienced snap-fit designers have learned the subject through a combination of intuition and trial-and-error. They may find the theory behind some of their knowledge in this section.

1.5.4 Design for Assembly Practitioners

Practitioners of design for manufacturing (DFM) and design for assembly (DFA) will be pleased to find that the ALC supports and is totally compatible with those design philosophies. The original motivation for creating the construct was support of DFM and DFA. A design for assembly practitioner interested in encouraging wise use of snap-fits should read Chapter 7 to understand how the snap-fit development process is compatible with general engineering and design for assembly practices and how it can be integrated into existing workshops.

1.6 Chapter Synopses

Brief descriptions of each chapter follow. Use them to plan your path through the book. Each chapter will conclude with a summary and a list of the most important ideas presented in the chapter. Refer to these end sections as a quick reminder of the chapter content or use them as an overview before reading the chapter.

- Chapter 2—The Attachment Level™ Construct is described in detail. Its organization is explained and the important terms and relationships are defined. This chapter is essential to understanding the ALC.
- Chapter 3—The physical features that hold one part to another are introduced. These are called *constraint features* because they constrain one part to the other. The two major classes of constraint feature are locators and locks. Inflexible "L" shaped features called lugs provide strength and are an example of a locator feature. The popular cantilever hook is a lock feature.

- Chapter 4—The idea of enhancements is introduced and various enhancement features are described. A snap-fit application only *requires* proper constraint, but we find that the best snap-fits show an attention to detail that goes far beyond just the constraint features. Enhancements are the kind of details an experienced designer may know to use but the novice will not.
- Chapter 5—Some important and fundamental concepts for understanding the behavior of snap-fits are explained. These include constraint and decoupling.
- Chapter 6—Feature level design and performance calculations are discussed. Some modifications to common feature calculations are suggested for increased accuracy. Some plastic materials principles related to feature analysis are briefly discussed. Rules of thumb for initial feature dimensions are provided.
- Chapter 7—A step-by-step process for developing a snap-fit attachment is presented. When applying the ALC to a snap-fit application, expect to refer to this chapter frequently until the development process becomes second nature. Cross-references to information in other chapters are provided.
- Chapter 8—A logical approach for diagnosing common snap-fit application problems is explained. Most snap-fit problems can be at least partially attributed to an attachment level cause. The basic premise is that feature level problems can not be addressed until attachment level shortcomings are fixed. Suggestions for approaching and fixing feature level problems are also provided.

1.7 Extending the ALC to Other Attachments

By the end of this book, it should be clear that, to ensure success with snap-fit applications, a systematic approach like the ALC should be used and the fundamental rules and requirements for snap-fits must be followed. It may not be as obvious that such an approach can also be quite useful for understanding and developing other kinds of attachments. This topic is discussed in more detail in Chapter 2.

1.8 Summary

Chapter 1 was an introduction to this book and to snap-fit technology. The idea of both a feature level and a systems or attachment level of snap-fit design was introduced. Some benefits of a systems approach to snap-fit development and design were discussed.

To employ an over-used but appropriate term, attachment level thinking is a snap-fit *paradigm shift*. It moves the snap-fit development process away from just the individual snap-fit feature. Instead, it provides a framework for thinking about the snap-fit as a *system* of interacting features and moves the snap-fit design process closer to end product considerations. By learning and applying the principles in this book, the reader will:

- Gain valuable insights into exactly how snap-fits work. An additional benefit is increased understanding of how all mechanical attachments work.
- Be able to design better, more effective snap-fit applications and be able to design them faster.
- Save product cost and support design for assembly by proper use of snap-fits.
- Learn how to think about snap-fits.

After studying some sophisticated snap-fit applications, one cannot help being impressed and maybe intimidated by their creativity and cleverness. It's OK to be impressed, but do not be intimidated. Personal experience is that very few, if any, really good snap-fit applications are designed that way in one sitting, particularly the more complex ones. Snap-fits involve a level of detail and creativity that generally requires evolution of the attachment into its final form. The development process described in Chapter 7 supports that evolutionary process and is likely to reduce the number of design iterations required. An important rule to remember is that good snap-fits are the result of *attention to detail*. Study any snap-fit applications, complex or simple, and you will find that the best ones always show a high level of attention to detail.

Next, Chapter 2 provides a detailed description of the Attachment Level™ Construct. With the help of the ALC, you will understand snap-fits well enough to create "world class" attachments yourself. A confidence building exercise to do as you learn about snap-fits is to critique them (on toys, interior trim on cars, household products, appliances, etc.) every chance you get. After a while, you will find yourself noticing how just about every application you study can be improved. Many of the improvements are no-cost; they are simply doing the right thing in the initial design. Again, a hands-on and a team approach to learning is highly recommended.

1.8.1 Important Points in Chapter 1

- The Attachment Level™ Construct (ALC) is a knowledge construct for explaining and organizing the concepts of snap-fit technology.
- Feature level aspects of snap-fit development are contained in the ALC.
- Experience with traditional mechanical methods of attachment (loose fasteners across an interface) is not suitable for developing snap-fit interfaces. New ways of thinking about function, component interfaces and attachments must be learned. Rivets, nuts, bolts and screws are not snap-fits; the knowledge does NOT transfer!
- Snap-fit attachment level knowledge, however, does transfer to other mechanical attachments. Applying attachment level principles can help improve development and design of all interfaces and support design for assembly.
- The root causes of most attachment problems in plastic parts are at the attachment level, not the feature level. Therefore, prevention, diagnosis and solution of application problems must start at the attachment level.

- Much of the attachment level development process will focus on developing a fundamentally sound snap-fit concept prior to beginning detailed math analysis.

References

1. Ary, D., Jacobs, L.C., Razavieh, A., *Introduction to Research in Education*, 5th Edition, (1996) p. 27–28. Orlando, Florida: Harcourt Brace College Publishers.
2. Boothroyd, G., *Design for Manufacture and Life-Cycle Costs* (1996), SAE Design for Manufacturability TOPTEC Conference, Nashville, TN.
3. Porter, C.A., Knight, W.A., *DFA for Assembly Quality Prediction during Early Product Design*, (1994). Proceedings of the 1994 International Forum on Design for Manufacture and Assembly, Newport, RI. Boothroyd Dewhurst, Inc., Wakefield, RI.
4. Deming, W.E., *Out of the Crisis*, (1982). p. 19. Cambridge, MA: Massachusetts Institute of Technology, Center for Advanced Engineering Study.
5. Goldsworthy, W.B., Heil, C., *Composite Structures are a Snap*, SAMPE Journal, (1998) v34 n1, pp. 24–30.
6. Lee, D.E., Hahn, H.T., *Composite Additive Locking Joint Elements (C-Locks) for Standard Structural Components*, Proceedings of the ASC Twelfth Annual Technical Conference, (1997) p. 351–360.
7. Lee, D.E., Hahn, H.T., *Assembly Modeling and Analysis of Integral Fit Joints for Composite Transportation Structures, 93-DETC/FAS-1362*, Proceedings of the 1996 ASME Design Engineering Technical Conferences, Irvine, CA.

2 Overview of the Attachment Level™ Construct

For the field of snap-fit attachment technology, the Attachment Level Construct (ALC) organizes the various relationships, concepts and rules for snap-fit attachments into a useful structure.

2.1 Introduction

The complete ALC for snap-fits is shown in Fig. 2.1. The construct contains three major groups: key requirements, elements and the development process. In this chapter, the key requirements and the elements will be introduced and explained. For completeness, the snap-fit development process is also shown although it is not discussed in detail until Chapter 7.

Key requirements are the common technical characteristics shared by all fundamentally sound snap-fits and they describe the important relationships between the *elements*. We know that specific application requirements (durability and ease of assembly, for example) cannot be efficiently or consistently met unless the snap-fit key requirements are satisfied. Because they are universal attachment requirements that must be satisfied, the key requirements describe the domain within which the snap-fit elements and development process exist.

Elements are either physical features of a snap-fit attachment or certain attributes used to describe or characterize the snap-fit application. *Constraint features* (locks and locators) and *enhancements* are physical elements of the attachment. The other elements are descriptive or spatial. The elements are used at specific times during the development process to make decisions about and to build the snap-fit interface.

A reminder: To make the terminology clear and to reinforce learning, find some products that use snap-fits and refer to them as you read. Identify the key requirements and elements as they are defined.

2.2 The Key Requirements

The key requirements are *strength, constraint, compatibility,* and *robustness*. They are a snap-fit's fundamental goals and they describe the desired relationships between the elements, Fig. 2.2. Because they are goals, satisfying the key requirements is the criteria for

2.2 The Key Requirements   15

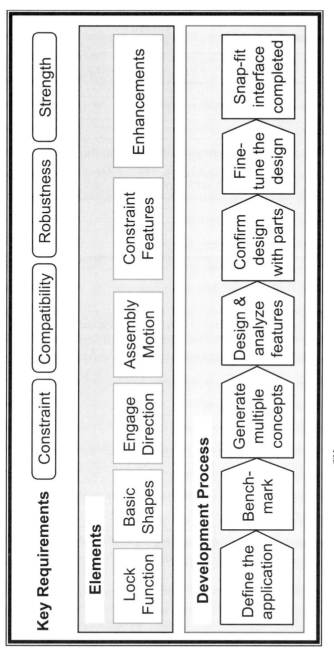

Figure 2.1 The Attachment Level™ Construct for snap-fits

16   Overview of the Attachment Level™ Construct

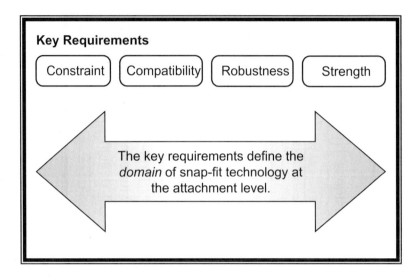

Figure 2.2 The snap-fit key requirements

judging the success of a snap-fit design. Using the key requirements and the elements, we will be able to describe the important attachment level design guidelines and rules. The following sections explain each of the key requirements in detail.

2.2.1   Strength

Strength is the performance of lock features during assembly and the ability of both lock and locator features to ensure attachment integrity for the life of the product. Attachment integrity means maintaining part-to-part constraint without looseness, breakage or squeaks and rattles. The product's useful life includes initial handling and assembly, operation (of a moveable snap-fit) and release and reassembly for maintenance or repair.

We should be familiar with strength because it is the basis for the traditional feature level approach to snap-fit design. Analytical methods for determining proper geometry and strengths of locators and locks are well documented. In Chapter 6, we discuss analytical methods for evaluating strength and assembly performance of snap-fit features.

In a snap-fit, as with most attachments, retention strength is generally the most important requirement. When we analyze snap-fit constraint features, we evaluate their performance and design them to ensure they are indeed strong enough to survive assembly, carry loads and resist forces. That is what we mean by feature strength. Strength, however is a component of a more global design requirement called reliability.

Reliability is the attachment's ability to hold parts together for the life of the product without failure. Reliability requires feature strength but it also requires the attachment to be properly assembled, used and serviced so that the designed-in strength is not lost. The attachment can fail when this second group of requirements is not met, not because of

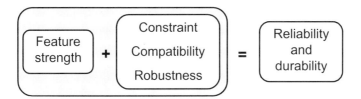

Figure 2.3 **Strength alone does not guarantee a good attachment**

inherent weakness, but because of improper assembly, use or service. So, the attachment must be more than strong, it must be reliable. Reliability is ensured when adequate feature strength is supplemented by the other three key requirements, Fig. 2.3.

Strength was described first because it is generally the ultimate goal of an attachment. Strength, however, is a potential and it cannot be achieved reliably or cost effectively unless the other three key requirements are met. The discussions of the remaining three key requirements will explain how they affect attachment strength.

2.2.2   Constraint

Constraint is prevention or control of relative movement between parts. In a snap-fit, locator and lock features provide constraint by transmitting forces across the interface and by positioning the mating and base parts relative to each other, Fig. 2.4.

In Fig. 2.4, we also introduce two generic snap-fit applications that will be used whenever possible to illustrate the concepts being discussed. They both represent relatively common types of snap-fit applications. The first, Fig. 2.4a, is a solid attaching to a surface. The second is a panel attaching to an opening, Fig. 2.4b. The terms *solid*, *surface*, *panel* and *opening* are four of the basic shapes used to describe snap-fits. The concept of basic shapes will be explained shortly.

All the key requirements are important to the attachment's performance and reliability, but constraint is the most fundamental requirement of a snap-fit. Success in the other three key requirements depends on a properly constrained snap-fit. Because it describes feature interactions, constraint is strongly tied to the idea of the snap-fit as a system.

Consider the mating part in a snap-fit as an object in space and the base part as ground. A free object in space can move in any of 12 ways. Six are translational movements (+ or −) along the three axes of a Cartesian coordinate system and six are rotational movements (+ or −) around the axes, Fig. 2.5. We will call these six linear and six translational movements *Degrees of Motion* or DOM. A totally unconstrained mating part can move in all 12 DOM relative to the base part. All 12 motions cannot occur simultaneously although combinations of rotation or translation involving any three adjacent axes are possible. The snap-fit's purpose is to prevent or control (i.e. constrain) mating part movements relative to the base part in all 12 DOM. Thus we can quantify constraint in terms of degrees of motion (DOM).

18   Overview of the Attachment Level™ Construct

(a) Solid to surface interface after assembly, (constraint features are hidden)

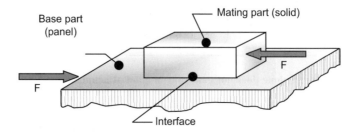

(b) Panel to opening, prior to addition of constraint features

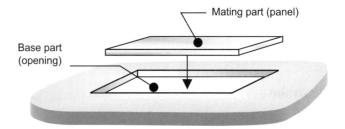

Figure 2.4   Constraint features in an attachment provide mating part to base part positioning and resist external forces

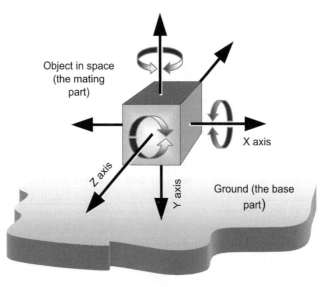

Figure 2.5   There are 12 possible directions or degrees of motion (DOM) for an object in space

Constraint features (locks and locators) appear in the snap-fit interface as constraint pairs, with a feature on one part engaging a feature on the other part. In most snap-fits, no relative movement is desired and the constraint pairs are arranged for constraint in exactly 12 DOM, Fig. 2.6a. In some snap-fits, however, relative motion between the joined parts is allowed and constraint may be less than 12 DOM as in Fig. 2.6b. However, the motion is controlled by the constraint features.

Design rule: In a fixed application, no relative motion between the parts is intended. The attachment is properly constrained when the mating part is constrained to the base part in exactly 12 DOM. In a moveable application, the attachment may be properly constrained in less than 12 DOM.

2.2.2.1 Improper Constraint

In any kind of snap-fit application, if constraint occurs in more than 12 DOM, then the application is over-constrained. If an application is constrained in less than 12 DOM (unless it is a moveable application), it is said to be under-constrained. Both under and over-constraint should always be avoided. Many snap-fit problems that appear to be caused by weak features are, in reality, the result of improper constraint. Table 2.1 shows some of the common problems that can occur with over or under-constraint conditions. The idea of constraint in the component interface is very important and many design rules apply to this requirement.

The subject of constraint is covered in detail in Chapter 5. Application of constraint principles during the snap-fit design process is discussed in Chapter 7.

(a) A push-button switch mounted to an opening is a fixed application

(b) A pulley wheel snapped to a bracket is a moveable application; the pulley can spin after it is snapped in place

Figure 2.6 Constraint features can either restrict all relative motions or they can control movement

Overview of the Attachment Level™ Construct

Table 2.1 Proper Constraint Versus Underconstraint and Overconstraint

Effect on	Constraint condition		
	Proper	Under	Over
Noise	Allows a close fit between parts	Part misalignment, possible looseness, squeaks and rattles	No direct effects
Assembly	Features fit without interference	No effects	Difficult assembly due to interference between features
Cost	Permits (cost-saving) normal or loose tolerances	No direct effects	Requires close tolerances
Analysis	Makes feature analysis possible	No effects	Interface is statically indeterminate
Reliability	Supports feature strength for reliability	Improper lock loading can lead to lock failure	Possible failure due to residual strain between constraint features. Possible component distortion under temperature extremes

2.2.3 Compatibility

Compatibility is harmony in the snap-fit interface between all the elements. It is the result of selecting the assembly motion and engage direction and arranging the constraint features to comprehend the components' basic shapes and allow ease of assembly. Some combinations of basic shapes, constraint features, assembly motions and engage directions are preferred; others can result in difficult assembly and/or feature damage and should be avoided. Incompatibility is often a subtle mistake, not easily recognized until symptoms and problems occur in assembly. This is why improved spatial awareness and reasoning is important in snap-fit development. We do not quantify compatibility as we do constraint; instead it is used as a factor in qualitative judgments about attachment options. Two examples of poor compatibility follow.

The first application shows assembly motion/constraint feature incompatibility, Fig. 2.7. The mating part has a lug, an inflexible locator feature. The wall on the right side of the base part restricts the available directions for the tip assembly motion required by presence of a lug. The location of the lug means the operator must try to force the lug to deflect enough to engage the hole in the base part. High assembly effort as well as broken parts are the most likely result.

(a) Solid to surface as assembled

(b) To assemble properly, the lug (2) must engage <u>before</u> the corner (1) makes contact

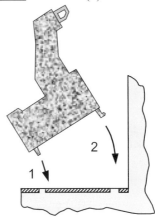

Figure 2.7 Assembly motion/constraint feature incompatibility where the design forces the features to engage out of order

The second example shows two instances of compatibility violations, Fig. 2.8. During assembly there is not enough clearance for the lock features to deflect. The result is higher assembly effort and immediate damage to either the retention faces of the hooks or the edges they engage. The second violation in this example is an assembly/disassembly motion incompatibility. The assembly motion is a push, but the finger-pull feature forces the disassembly motion to be a tip. This causes over-deflection damage to the hooks at the finger-pull end of the panel and possible damage to the locator pins at both ends of the panel.

Important compatibility rules are:

- All physical features in the interface must be compatible with the assembly motion.
- The selected assembly motion must be compatible with the basic shapes.
- The assembly and disassembly motions should be the same (although opposite in direction).
- Allow clearance for feature deflection during assembly and disassembly.

These are simple and seemingly obvious rules, yet they are violated. Both of the examples of compatibility violations shown here are based on actual applications.

2.2.4 Robustness

Robustness is often defined as tolerance to dimensional variation; as a snap-fit requirement, it has a broader meaning. We define snap-fit robustness as tolerance of the snap-fit to all the variables and unknowns that exist in product design, manufacture, assembly and use. Robustness is indeed tolerance to variation, but that variation is caused by many unknowns

22 Overview of the Attachment Level™ Construct

(a) The application is a solid attaching to an opening

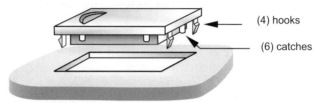

(b) Assembly is a push motion and the design has inadequate clearance for hook deflection

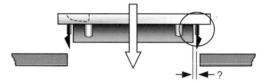

(c) As assembled

(d) Disassembly requires a tip motion causing over-deflection damage to the hooks at one end and possible damage to other constraint features

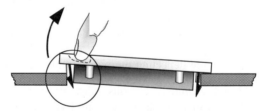

Figure 2.8 Compatibility violations in a simple application

and manifests itself in many disagreeable ways. Unknowns in the life of a snap-fit can include a wide variety of situations, including:

- The customer's ability to interpret how to use or operate the snap-fit.
- A service technician's ability to disassemble and reassemble the attachment without damage.

- The working environment and conditions in which the parts are assembled.
- The possibility of misuse, unexpected loads.

An example of the importance of robustness and its relation to strength in a snap-fit is appropriate here. We will use a very simple application, represented in Fig. 2.9 by the basic shapes *panel* and *opening*. This example is based on an actual design problem and we will be referring to this particular application again in later chapters. For our purposes now, it is sufficient to summarize the problem and solution very briefly.

The locking features in the original design are four cantilever hooks, one at each corner of the mating part. The panel is a very low mass part and no external forces are applied to it once it is in place. Each hook had been analyzed to ensure adequate strength for both assembly and long-term retention of the panel to the opening. However, in spite of sufficient strength in the hooks, some panels were falling off within the first few months of use. Investigation of failed parts showed damage to one or more of the hooks. Some had taken a permanent set while other hooks were broken off completely.

A first reaction to this problem might have been to simply strengthen the hooks; after all they were breaking. This is a feature level fix and, as we will see, would have been a mistake. A very important rule for understanding and fixing snap-fit problems is that *feature level problems cannot be fixed until we verify there are no attachment level problems in the snap-fit*. A study of the assembly process for this part revealed that:

- The application was in a vertical plane below the assembly operator's natural line of sight.
- It was a blind assembly, the operator's hand would hide the attachment area as they held the mating part (the panel) and tried to position it in the opening.
- The operator's fingertips, as they grasped the panel in a normal manner, would contact the area around the opening before the locks were properly positioned around the edge of the opening, Fig. 2.10.

After observing the assembly operation, it was not too hard to conclude that the root cause of the part problem was damage to the hooks during the assembly operation. Making the hooks even stronger might have prevented damage but would have also increased the assembly force, which could have caused ergonomic problems; and there were no guarantees the problem would be solved. Other problems that could have also been occurring, which stronger hooks would certainly not have solved, were the extra time it took the operator to

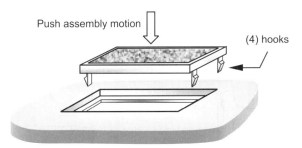

Figure 2.9 **The application is a small panel attaching to a recessed opening**

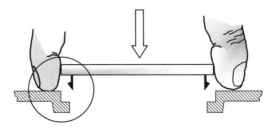

Figure 2.10 The operator's fingertips interfere with proper hook alignment and engagement

finesse the panel into place for assembly and the continuous frustration of carrying out a difficult assembly.

What if the parts had been assembled in an automated operation, such as with a robot? Higher assembly forces and frustration would no longer be an issue, but the required precise positioning of the panel to the opening might have caused problems, even for an automated operation.

One possible fix for this problem is shown in Fig. 2.11. With the addition of pins that serve as both *locators* (constraint features) and as *guides* (an enhancement feature), the application is now robust to the mechanics of the assembly process. The pins, first acting as guides, engage the edge of the opening to orient and stabilize the mating part before the operators' fingers contact the base part. This ensures the hooks are in proper position for engaging the edge and will not be damaged. The operator can easily position the panel in the opening with the hooks resting against the edge and, with a final push, engage the hooks to complete the assembly. Note that in this application, there was sufficient clearance for the long pins proposed as a fix. However, other solutions were available if the clearance had not been available. They will be discussed when we revisit this particular application in Chapter 4.

This was a rather simple problem to identify; the real issue is why it happened in the first place. Possibly the designer considered a panel to an opening as too basic and simple to worry about. Thinking about the attachment as a system could have prevented operator frustration, customer dissatisfaction, product warranty costs, engineering time spent to change the design and tooling costs to modify the mold.

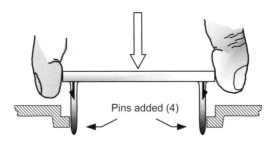

Figure 2.11 Possible fix to stabilize mating part and prevent hook damage during assembly

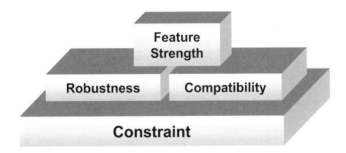

Figure 2.12 **The four key requirements**

Another important lesson to take away from this example: Never try to understand a snap-fit problem without first watching it being assembled. If possible, perform the assembly operation yourself to get a real feeling for what is happening.

To summarize, the hooks in this example had enough *strength* to survive normal assembly deflections and to hold the panel in place once it was engaged. But, strength was not enough. The entire system was not *robust* to the assembly process; therefore the attachment was not reliable. In conclusion, robustness helps to ensure that feature strength is properly utilized; this in turn ensures reliability of the snap-fit attachment.

As with compatibility, we do not quantify robustness. But it is an important goal and, as such, it should influence many design decisions. The enhancements described in Chapter 4 address many robustness issues.

This completes the discussion of the four key requirements. Their relationship is shown in Fig. 2.12. The ultimate goal is feature strength for attachment durability and reliability. Proper constraint is the most fundamental of the requirements and is the basis for the other three requirements. Robustness and compatibility depend on proper constraint to be effective and also help to enable feature strength.

2.3 Elements of a Snap-Fit

Six elements make up the descriptive/spatial and physical description of a snap-fit attachment. By learning them, you will be building an organized structure of snap-fit technology in your mind. This is something like defining folders in a filing system. With this "filing system" you will find that remembering and using specific snap-fit features in design solutions will be made easier.

The six elements are divided into two groups, Fig. 2.13. Four *descriptive/spatial* elements are used to describe the application in specific attachment level terms that will help us apply the development process. Two *physical* elements are used to describe the actual features (or building blocks) of the attachment interface.

26 Overview of the Attachment Level™ Construct

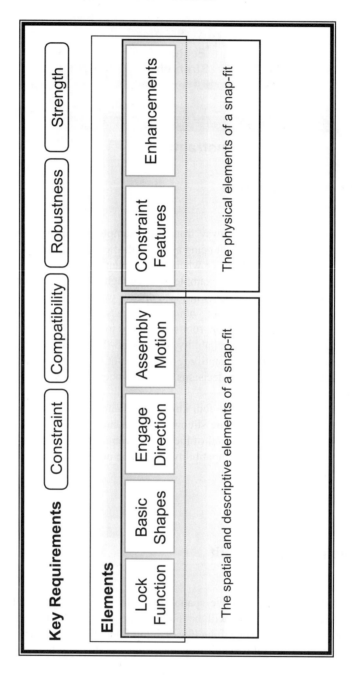

Figure 2.13 The six elements of a snap-fit

2.3 Elements of a Snap-Fit   27

2.3.1   Function

Function is the first of the descriptive elements. It is the attachment's fundamental purpose, what the locking features in the snap-fit must do. Function is not one of the more important elements in terms of developing a snap-fit. However, it is useful in grouping lock features with respect to various performance requirements, thus it contributes to an overall understanding of the snap-fit technology. Function is described in terms of *action*, *attachment type*, *retention* and *lock type*:

2.3.1.1   Action

Action is the potential for movement designed into the snap-fit application.

In *fixed* snap-fits, no relative motion between parts can occur after they are locked together. The application is constrained in exactly 12 degrees of motion. The push-button switch in Fig. 2.6a and the panel-to-opening examples in Figures 2.8 and 2.9 are also fixed snap-fits.

In *moveable* snap-fits, relative motion can occur between the joined components when they are engaged. The components are never completely separated during this motion. When no constraint features limit this motion, it is free movement. The pulley shown in Fig. 2.6b is an example. When locks or locators control or regulate the motion so the mating part is *sometimes* immobile, it is controlled movement, Fig. 2.14.

Where free movement can occur, then no constraint exists in those directions and the attachment will be (properly) constrained in less than 12 degrees of motion. The pulley is properly constrained in 10 degrees of motion.

2.3.1.2   Attachment type

The snap-fit may be the final attachment or it may be temporary until some other attachment occurs.

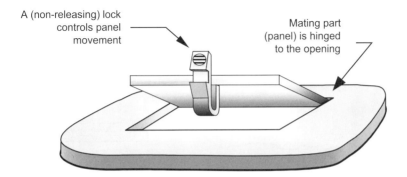

Figure 2.14   Controlled movement in a panel to opening

The snap-fit is *final* when it is the attaching method that will hold the application together throughout its useful life. Most snap-fits fall into this group and, in all the examples shown thus far, the lock is intended to be the final attachment.

Temporary snap-fits hold the application only until some other attachment occurs. They only need to be strong and effective enough to position the mating part to the base part until the final attachment is made. Temporary snap-fits can support design for assembly by allowing build-up of several parts prior to final attachment. They may sometimes save money by allowing a less expensive final attaching process to be used, a slow-cure instead of rapid-cure adhesive for example.

2.3.1.3 Retention

Retention refers to the nature of the locking pair: permanent or non-permanent.

Permanent locks are not intended for release, Fig. 2.15. No lock is truly permanent, but these locks, once engaged, are difficult to separate. In some cases, they can be released with tools or high effort, but damage to the lock or parts may result. They are indicated where non-serviceable attachments are to be made or where evidence of product tampering is required. They may also be useful where an attachment must survive sudden impact forces that could cause a non-permanent lock to release. Figure 2.15a shows a trap lock where the locking fingers are contained within the interface with no access for releasing them. Figure

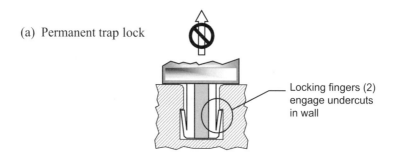

(a) Permanent trap lock

Locking fingers (2) engage undercuts in wall

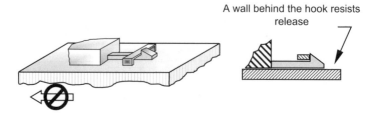

(b) Permanent hook lock

A wall behind the hook resists release

Figure 2.15 Permanent locks

2.15b is a hook engaging a strap-like feature on a wall. Assembly forces are high, but once engaged, the wall prevents hook end rotation for release.

Non-permanent locks are intended for release. Two kinds of non-permanent locks are identified in the *lock type* classification.

2.3.1.4 Lock type

Refers to how the lock feature works to allow part separation, Fig. 2.16.

Releasing locks are designed to allow part separation when a predetermined separation force is applied to the parts, Fig. 2.16a.

Non-releasing locks require manual lock deflection for part separation, Fig. 2.16b. A non-releasing lock is also shown in Fig. 2.14. Note that non-releasing locks may release under certain conditions. The non-release function is not a guarantee against unintended separation.

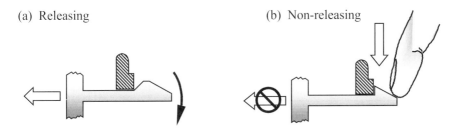

Figure 2.16 Non-permanent locks

2.3.1.5 Function Summary

Function describes exactly what the locking features must do in the application. As we will see in Chapter 3, some lock features are better than others at performing some of these functions. The function decision is summarized in Fig. 2.17. Describe an application by starting at the top and working down through the four levels. Table 2.2 lists examples of the various ways snap-fits are classified by function.

2.3.2 Basic Shapes

Basic shapes are the second descriptive element. They are simple geometric shapes that describe the parts being attached. Classifying components by shape allows us to think of an application in generic terms. This is important because it helps us transfer snap-fit concepts between applications. Most importantly, however, use of basic shapes helps us to visualize the attachment. This supports the spatial reasoning needed to develop good snap-fit concepts.

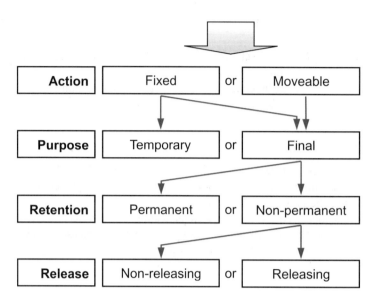

Figure 2.17 Lock function flowchart

Table 2.2 Function Examples

Application	Action	Purpose	Retention	Release
Switch assembly into an opening. Similar to Fig. 2.6a.	Fixed	Final	Generally non-permanent for service	Releasing if no access for manual release
The rocker switch in a switch assembly.	Moveable (controlled)	Final	Generally non-permanent for service	Releasing or non-releasing
Battery access panel (slide to release) in a TV remote control unit.	Fixed	Final	Non-permanent	Generally releasing
Battery access panel (tip to release) in a toy. Similar to Fig. 2.14.	Fixed	Final	Non-permanent	Generally non-releasing
Access cover for a circuit board requiring manufacturer service only.	Fixed	Final	Permanent	N/A
Pulley to a bracket, Fig. 2.6b.	Moveable (free)	Final	Generally non-permanent for service	Releasing or non-releasing
Lamp lens snapped to lens carrier prior to epoxy bonding.	Fixed	Temporary	N/A	N/A

2.3 Elements of a Snap-Fit

Table 2.3 Basic Shapes Summary

Part	Basic Shapes					
	Solid	Panel	Enclosure	Surface	Opening	Cavity
Mating part	Common	Common	Common	Rare	Rare	Low
Base part	Common	Rare	Rare	Common	Common	Common

2.3.2.1 Mating Part and Base Part

We start describing basic shapes by defining the two components that make up a typical snap-fit as the *mating part* and the *base part*.

The base part may be large and obviously stationary or fixed. The mating part is typically smaller than the base part, held in the hand(s) and moved into attachment with the larger, stationary base part. The push button switch and the pulley in Fig. 2.6, the generic solids in Figures 2.7 and 2.8, and the small panel in Fig. 2.9 are all considered mating parts. The mating part will generally be one of the three basic shapes: *solid*, *panel* or *enclosure*. The base part will generally be a *solid*, *surface*, *opening* or *cavity*, Table 2.3.

We can usually identify the mating and base parts by using the size and movement criteria described above. We can also use the basic shape for identification. If all these fail to distinguish the mating from the base part, then the parts are probably so similar that an arbitrary selection can be made. Note that these distinctions are true *most of the time*, some exceptions will be shown later in this section. Exceptions do occur, but that does not reduce the value of having these definitions.

2.3.2.2 Basic Shape Descriptions

- Solid—Components with both rigidity and depth, Fig. 2.18. Solids may have constraint features in three dimensions.
- Panels—Relatively thin components, they tend to be compliant in bending and torsion, Fig. 2.19. Constraint features are generally at or near the perimeter but can be anywhere on the panel.

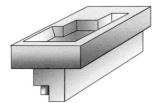

Figure 2.18 Solids

32   Overview of the Attachment Level™ Construct

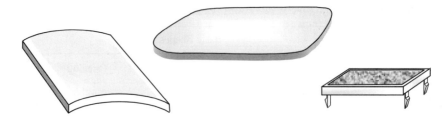

Figure 2.19   Panels

- Enclosure—A three-dimensional cover, Fig. 2.20. An enclosure is essentially a three-dimensional panel. They have compliant walls and constraint features along the open edges.
- Surface—A locally two-dimensional area, Fig. 2.21, with constraint features located on the surface. Note that while a panel is not normally a base part, a surface on a panel could be a base part.
- Opening—A hole in a surface, Fig. 2.22, with the constraint features located at or near the edges of the opening. Again, while a panel is not a base part, an *opening* in a panel is a base part. See Figures 2.8 and 2.9.
- Cavity—A cavity is an opening with depth, Fig. 2.23. Constraint features will occur in three dimensions.

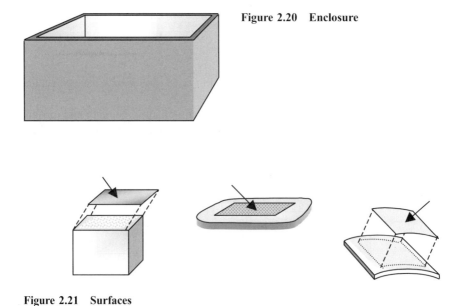

Figure 2.20   Enclosure

Figure 2.21   Surfaces

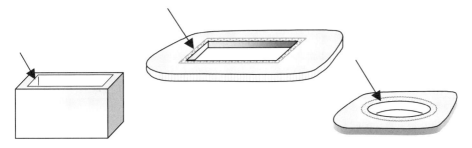

Figure 2.22 Openings

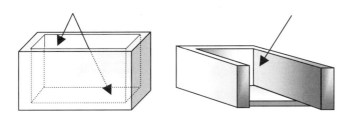

Figure 2.23 Cavities

2.3.2.3 Basic Shape Summary

Using generic descriptions of part shapes helps us transfer important snap-fit knowledge, past experience and lessons-learned between applications. For example, a panel to an opening application may be a small closeout panel or reflector on a cabinet, a speaker grille in an automobile interior or a larger door or access panel. Regardless of the application, the *fundamental* design principles for a panel to opening snap-fit application will always be true. By learning about a limited number of basic shape combinations, we will be learning about the most common product applications.

Table 2.4 shows how the basic shapes are commonly distributed between the mating and the base part. Consideration of the most common and geometrically possible combinations leads to the summary in Table 2.5. These tables are based on review of hundreds of applications, most of them automotive. Reviews of other products, however, seem to be in general agreement with these observations. The judgements of frequency are subject to change as more information is gathered; comments and examples from interested readers are always welcome.

The value in having these tables is that we can begin to classify and group our applications according to their shapes. Design knowledge can readily transfer between applications that fall within one cell. Certain knowledge will also transfer within a row or a column or between cells with shapes having similar characteristics. For example:

- Enclosure/panel shapes where an enclosure is defined as having walls resembling the panel shape,

Table 2.4 Observed Frequencies of Basic Shape Combinations

Mating part shapes	Base part shapes					
	SOLid (common)	**PAN**el (rare)	**ENC**losure (rare)	**SUR**face (common)	**Op**ening (common)	**CAV**ity (common)
SOLid (common)	SOL-SOL high	C	SOL-ENC Low	SOL-SUR high	SOL-OP high	SOL-CAV high
PANel (common)	PAN-SOL low	PAN-PAN low	PAN-ENC Low	PAN-SUR low	PAN-OP high	PAN-CAV low
ENClosure (common)	ENC-SOL low	C	ENC-ENC Low	ENC-SUR high	ENC-OP low	ENC-CAV low
SURface (rare)	SUR-SOL low	C	SUR-ENC Low	C	X	X
Opening (rare)	X	X	X	X	X	X
CAVity (low)	X	X	X	X	X	X

High—A very common basic shape combination.
Low—Less frequently observed.
C—Covered by some other combination. (Subject to change).
X—Judged to be geometrically impossible. (Subject to change).

Table 2.5 Most Common Basic Shape Combinations; the High Usage Area of Table 2.4

Mating part shapes	Base part shapes				
	SOLid (common)	**ENC**losure (rare)	**SUR**face (common)	**Op**ening (common)	**CAV**ity (common)
SOLid (common)	SOL-SOL high	SOL-ENC low	SOL-SUR high	SOL-OP high	SOL-CAV high
PANel (common)	PAN-SOL low	PAN-ENC low	PAN-SUR low	PAN-OP high	PAN-CAV low
ENClosure (common)	ENC-SOL low	ENC-ENC low	ENC-SUR high	ENC-OP low	ENC-CAV low

High—A very common basic shape combination.
Low—Less frequently observed.

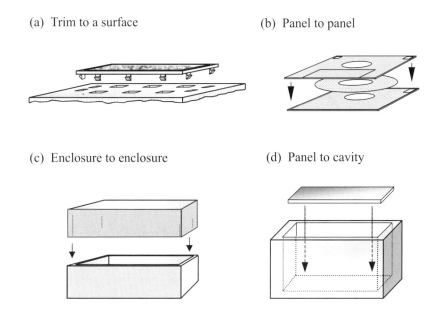

(a) Trim to a surface (b) Panel to panel

(c) Enclosure to enclosure (d) Panel to cavity

Figure 2.24   Less common basic shape combinations

- Opening/cavity shapes where the cavity can be thought of as an opening with depth.
- Surface/panel shapes when the surface is located on a panel.

Certain basic shape combinations have good and bad characteristics. Some basic shape combinations should be avoided. Each combination can have certain preferred assembly motions, constraint features and enhancements that help ensure a good snap-fit. Once these other elements of a snap-fit are discussed, we will be able to summarize some desirable and undesirable characteristics for the common basic shape combinations.

The combinations in Table 2.5 are the most common. Some exceptions to the general rules are shown in Fig. 2.24. A badge or a covering trim application would be a panel to a surface, Fig. 2.24a. A computer diskette cover assembly, Fig. 2.24b, is a panel to panel application, although we can also think of it as surface to surface. We can also imagine an enclosure to enclosure application, Fig. 2.24c, and a panel to a cavity, Fig. 2.24d.

2.3.3   Engage Direction

Engage direction is the third descriptive element. It is the *final* direction the mating part moves as it locks to the base part and is described by a directional vector (of zero magnitude) defining the mating part's movement as locking occurs, Fig. 2.25a. Note that there may be movements of the mating part in space prior to the final engaging motion, the direction(s) of those movements is not considered engage direction. As we select an engage direction, we are also, by default, selecting a separation direction; it is the relationship of the separation

(a) Engagement is in the -Y direction

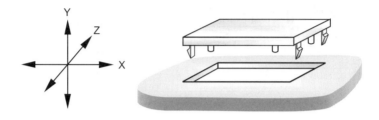

(b) The lock features engage in the -Y direction and separate in the +Y direction.

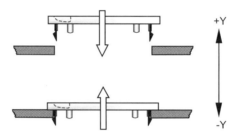

Figure 2.25 Engage direction

direction and the locking features we are most concerned about, Fig. 2.25b. The locking features (lock pairs) will be required to resist any forces on the attachment that tend to separate the parts and, generally, the locking feature(s) is the weak link in the attachment system. An important rule when identifying allowable engage directions is:

Select an engage direction so that the (opposite) separation direction is not in the same direction as any significant forces on the attachment.

This simple rule means that there should be no significant transient or long-term forces trying to release the lock and separate the parts. In Fig. 2.26a we see a solid to opening application having two available engage directions. The preferred engage direction is in the −Y direction so that the separation direction is *opposite* the force on the mating part. This allows the force to be carried by the surface of the flange (a locating area) on the solid rather than by any locking features.

What is a significant transient force; what is a significant long-term force? That is up to the designer to determine with the help of a polymers expert. The answer will depend on the magnitude of the force, the force history and the long and short-term performance

2.3 Elements of a Snap-Fit   37

(a) Two possible engage directions (+Y and -Y) for the solid to opening application

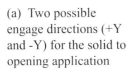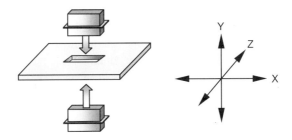

(b) Select the engage direction (-Y) that is in the same direction as the force on the mating part

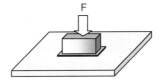

Figure 2.26   Forces on the mating part should be resisted by locators, not locks

characteristics of the material chosen for the part. Sometimes, a significant force turns out to be an unexpected force due to misuse of the product or accidental impact.

When the application is such that significant loads can occur in the lock direction, there are sometimes steps that can be taken to ensure that the lock features will not release. Changing to a different lock style is one. Making the lock permanent (as defined under function) is another. However, simply making a lock feature non-releasing *will not* guarantee against unintended release. Some of the performance enhancements (Chapter 4) can help improve lock retention strength. The decoupling principles described in Chapter 5 explain how some lock styles can have more retention strength than others. In Chapter 3, lock strength and lock behavior are discussed in detail.

Sometimes, presence of a significant force in the removal direction is a reasonable argument for not using a snap-fit attachment and, instead, considering other fastening methods.

Note that engage direction refers to the lock pair's primary movement as it engages. This is the in the same direction as the mating part. Other movements will occur in one or both lock pair members as they deflect to allow engagement, Fig. 2.27.

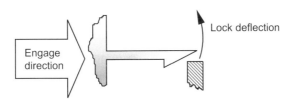

Figure 2.27   Lock engage direction is not the same as lock deflection

While there may be a number of *possible* engage directions, the truly feasible engage directions for any particular application are limited. In addition to the limitations imposed by external forces, other limitations can be interactions between the parts' basic shapes, ergonomics, packaging and access conditions.

2.3.4 Assembly Motion

Assembly motion is the fourth and last descriptive element. It is defined by the generic motions: *push*, *slide*, *tip*, *twist* and *pivot*. Think of assembly motion as what a human operator must do to assemble the components. It is the *final* motion of the mating part as it locks to the base part, Fig. 2.28. Assembly motion helps the designer visualize the mating-part to base-part assembly process. Like basic shapes, assembly motions support generic snap-fit descriptions and spatial reasoning for snap-fit concept development. They may also have ergonomic implications in some applications where an awkward position and excessive assembly force, when combined with a certain assembly motion, can result in increased likelihood of repetitive motion injury. As we will learn, they also have significant impact on the attachment design with respect to strength.

- Push—A linear movement where contact between the mating and base parts occurs (relatively) shortly before final locking, Fig. 2.28a. Some guide feature contact may occur before the locators or locks engage.
- Slide—A linear movement with early contact between locator pairs followed by additional mating part movement with continuous contact with the base part prior to final locking, Fig. 2.28b. Both push and slide are simple motions. The next three are more complex.
- Tip—A rotational movement. A locating feature(s) on the mating part, (1) in Fig. 2.28c, is first engaged to the base part. Initial engagement is followed by mating part rotation (2) around the initial locator pair until locking feature engagement occurs.
- Twist—A rotational movement. A mating part with axisymmetric constraint features is first engaged to the base part with a linear motion, (1) in Fig. 2.28d. The mating part is rotated (2) around the axis so its constraint features engage a complementary arrangement of constraint features on the base part. The behavior is similar to that of a "quarter-turn" fastener.
- Pivot—A rotational movement. The mating part is first engaged to the base part at one locator pair with a push motion, (1) in Fig. 2.28e. The mating part is pivoted around that point with continuous contact until lock engagement occurs (2). A pivot can be thought of as a combination of both the tip and slide motions, with rotation about one locator pair and continuous contact occurring simultaneously.

Note how some assembly motions may be preferable to others depending on the basic shapes involved, application accessibility and operator ergonomics. Table 2.6 shows some of these possibilities and provides an indication of the more preferred motions. We will also see how assembly motion contributes to creativity during the development process (Chapter 7) and, like basic shapes, helps organize thinking about applications.

(a) Push - panel to opening and solid to cavity

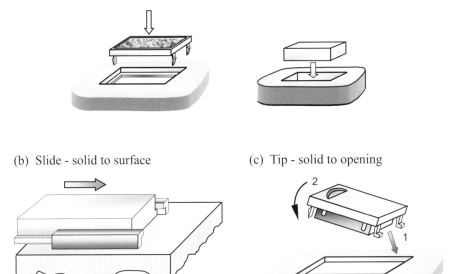

(b) Slide - solid to surface

(c) Tip - solid to opening

(d) Twist - solid to cavity

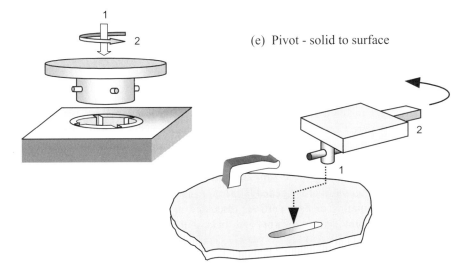

(e) Pivot - solid to surface

Figure 2.28   Assembly motions

Table 2.6 Common Basic Shape Combinations and Available Assembly Motions

Mating part shapes	Base part shapes				
	Solid	Enclosure	Surface	Opening	Cavity
Solid	Push Slide Tip Twist Pivot	N/A	Push Slide Tip Twist Pivot	Push Tip Twist*	Slide Tip* Twist
Panel	N/A	N/A	Push Slide Tip	Push Tip	Push Tip
Enclosure	Slide Tip Twist*	Tip	Push Tip Twist*	Push Tip Twist*	N/A

* Some availability, depending on specific part geometry. Twist is generally not preferred for large parts.

We are finished introducing the four *descriptive* elements of snap-fit design: function, basic shape, engage direction and assembly motion. They will be applied during the development process described in Chapter 7. We now move on to the physical elements. These are the actual "building blocks" of the snap-fit. Unlike the preceding detailed discussion of the descriptive/spatial elements, we will only provide a brief introduction to the physical elements here because they are discussed in great detail in the next two chapters.

2.3.5 Constraint Features

We have already explained that constraint involves controlling mating part movement relative to the base part. Constraint features are the mechanisms that provide constraint in the attachment. There are two kinds of constraint feature: *locator features* and *lock features*. Usually, the names are shortened to just *locators* and *locks*.

Locators and locks are the "necessary and sufficient" features for a snap-fit attachment. In other words, they are all that is needed to create a snap-fit. Both types of features can be found on either the mating or base part. As discussed in the section on key requirements, proper constraint as provided by locators and locks is the basis for a successful snap-fit.

2.3.5.1 Locator Features

Locator features (*locators*) are relatively inflexible constraint features, Fig. 2.29. They provide strength against forces across the interface and they provide precise positioning of the mating part to the base part. A good term that describes what locating features do is *nesting*. Think of locators as the interface features that cause the parts to nest together.

(a) Lugs and lands are distinct locator features

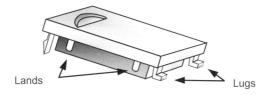

(b) Natural locator features

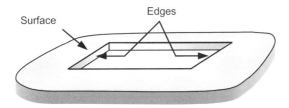

Figure 2.29 Locator constraint features

Locators may be distinct features added to the attachment strictly to provide locating, Fig. 2.29a, or they may be *natural locators*: pre-existing features on the mating or base part such as a wall, surface or edge that perform a locating function, Fig. 2.29b.

In fixed applications, locators prevent motion and carry loads in all but the mating part removal direction. In moveable applications, they may also be used to control or limit motion in the direction(s) of movement (controlled action).

Locator features are grouped into common types: *pin*, *cone*, *track*, *wedge*, *catch*, *surface*, *edge*, *lug*, *land*, *slot* and *hole*. We also classify the living hinge as a locator. The features are grouped in this manner because each type has a unique set of constraint behaviors.

Presence of a locator on one component implies a mating locator on the other. Together they make up a *locator pair*, Fig. 2.30. When we discuss locators in a snap-fit, remember that we are really referring to a locator pair. Locators are discussed in detail in Chapter 3.

2.3.5.2 Lock Features

Lock features, or simply *locks*, are constraint features which hold parts in the located or nested condition. With certain notable exceptions (discussed in Chapter 3), they are weak compared to locators because locks must deflect to allow assembly. Once the mating and base parts are located the locking features hold them in that position, Fig. 2.31, so the strong locators can do their job of providing positioning and carrying forces across the interface.

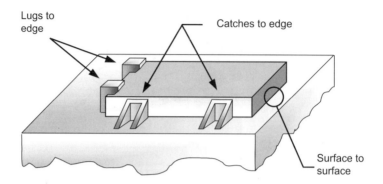

Figure 2.30 Locator pairs

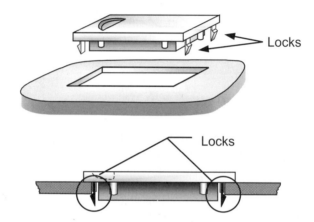

Figure 2.31 Once the parts are located, locks hold them in place

Integral locks are grouped into common types: hook, catch, annular, torsional and trap. They are defined in a particular application by specific functional characteristics. These characteristics were introduced when we defined the descriptive element *function*.

Because locks deflect elastically to allow assembly, they must be flexible (weak) in that direction. After deflecting for assembly, they return toward their initial position. This results in interference between the lock and the other half of the lock pair (located on the other part). As long as this interference is maintained, the parts are locked together. Because locks prevent mating part movement away from the base part, (the separation direction) they must have some strength in that direction, Fig. 2.32.

Presence of a lock on one component implies a mating feature on the other. Generally, the mating feature to the lock is a locator, not another lock, because the lock should engage a strong and inflexible feature. Together the lock and locator make up a *lock pair*, Fig. 2.33. As with locator pairs, when we discuss locking in a snap-fit, we always assume a lock pair exists. Locks are discussed in detail in Chapter 3.

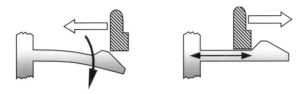

Figure 2.32 Locks must be both weak for engagement and strong for retention

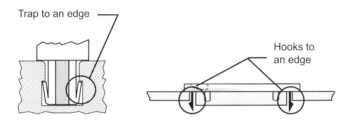

Figure 2.33 Lock pairs

2.3.6 Enhancements

Enhancements are the second group of physical elements. They can be separate and distinct interface features or they can be attributes of constraint features or other part features. Enhancements are a relatively undocumented aspect of snap-fit design; they are often the minor details that designers learn about through trial-and-error and, sometimes costly, experience. By being aware of and considering enhancement requirements during the initial development stages, the snap-fit designer can prevent both minor and major snap-fit problems. Some of these problems can be important enough to force redesign when a product fails early testing or has problems with performance in service. This, of course, can be expensive, time-consuming and embarrassing. Other problems will be minor irritations to the manufacturer, the assembler or the customer that will not seem important enough to force design changes, but they can increase costs, affect quality and productivity and reduce customer satisfaction.

Enhancements improve an attachment's robustness to variables and conditions encountered during the product's life. They can also improve user-friendliness. They do not directly affect the attachment's strength, but they can have important indirect effects on reliability. Enhancements often go unnoticed and unappreciated but they help make a snap-fit "world class". They are sometimes tricks-of-the-trade that experienced designers have learned about through experience. A novice, however, may not recognize the need for enhancements. Knowing about the different kinds of enhancements will also enable the designer to better study and interpret other snap-fit applications during product benchmarking.

Enhancements are classified into four major groups: *assembly, activation, performance* and *manufacturing*. Enhancements are described in detail in Chapter 4; here we will briefly introduce them:

Enhancements for Assembly—Features or attributes that support product assembly:

- Guidance—Ensures smooth engagement and latching of mating parts. Guidance enhancements are further broken down into *guides, clearance* and *pilots*.
- Operator feedback—Attributes and features ensuring clear and consistent feedback that the attachment has been properly made.

Activation enhancements—Informational and mechanical enablers that support attachment disassembly or usage:

- Visuals—Provide information about attachment operation or disassembly.
- Assists—Provide a means for manual deflection of non-releasing locks.
- User feel—Attributes and features that ensure a good feel in a moveable snap-fit.

Performance enhancements—Ensure that the snap-fit attachment performs as expected:

- Guards—Protect sensitive lock features from damage.
- Retainers—Provide local strength and improve lock performance.
- Compliance—Attributes and features that take up tolerance and help maintain a close fit between mating parts without violating constraint requirements.
- Back-up lock—Provides a back-up means of attachment.

Manufacturing enhancements—Techniques that support part and mold development, manufacturing and part consistency.

Many manufacturing enhancements are documented in standard design and manufacturing practices for injection-molded parts and are already recognized as important factors in plastic part design. They fit neatly into the ALC as enhancements and, because of their importance, are included.

- Process-friendly design—Following recommended and preferred plastic part design practices.
- Fine-tuning—Practices that allow for easy mold adjustments and part changes or fine-tuning.

Enhancements are summarized in Table 2.7 and are discussed in detail in Chapter 4.

2.3.7 Elements Summary

This concludes the elements of snap-fit design. Four of the elements are spatial or descriptive: *lock function, basic shape, engage direction* and *assembly motion*. Two elements are physical parts of the snap-fit: *constraint features* (consisting of locators and locks) and *enhancements*. The purpose of Chapter 2 has been to provide a detailed discussion of the spatial/descriptive elements and an overview of the physical elements. The physical elements require much more detailed explanation and are covered in Chapters 3 and 4. All of the elements will be applied during the snap-fit development process described in Chapter 7.

Table 2.7 Enhancements Summary

	Why	What
For assembly		
Guidance	Ease of assembly	Guide—stabilize parts Clearance—no interference Pilot—correct orientation
Feedback	Indicate good assembly	Tactile, audible or visual
For activation		
Visuals	Indicate disassembly, assembly and operation	Words, arrows, symbols
Assists	Enable disassembly, assembly and operation	Extensions for fingers or tools
User feel	Perceived quality	Force-time signature
For strength and performance		
Guards	Protect weak or sensitive features	Prevent over-deflection Reduce strain
Retainers	Strengthen locks	Increase retention strength Stiffen the lock area Support the lock
Compliance	Take up tolerances and prevent noise	Elastic features Local yield
Back-up lock	A back-up attaching system	Available fasteners Adaptable interfaces
For manufacturing		
Process-friendly	Consistent features	Simple designs
	Minimum cycle times	Follow mold and product design guidelines
Fine-tuning	Speeds development	Local adjustments
	Easy part fine-tuning and adjustments for quality	Metal-safe designs Adjustable inserts

2.4 Summary

The key requirements and elements of the Attachment Level™ Construct for snap-fits, Fig. 2.1, were described in Chapter 2. To support the human mind's desire for organization, snap-fit design technology has been described in terms of key requirements and elements. Key requirements are the common and fundamental goals of all good snap-fit attachments. Elements are the spatial/descriptive and physical parts of the domain that we use to make decisions about the snap-fit and construct the attachment concept. Using the key requirements and elements, we can also develop attachment level design guidelines and rules.

Some of the elements are generic, allowing the snap-fit designer to think in terms of simple shapes and motions. This supports the spatial understanding and reasoning that is so important to successful snap-fit design. It also enables transfer of useful snap-fit knowledge between applications. Specific meanings for snap-fit terms also allow clear, unambiguous communication between designers about snap-fits.

2.4.1   Important Points in Chapter 2

- The ALC defines and organizes the design space for snap-fits, explaining it in terms of key requirements, elements and a development process.
- Every snap-fit should satisfy the four key requirements: *constraint, compatibility, robustness* and *strength*.
- Feature strength for attachment reliability is the ultimate goal of most snap-fits and is one of the key requirements. To have reliable strength, a snap-fit must satisfy the other three key requirements.
- Constraint is the most fundamental of the key requirements. Proper constraint is required for success in the other requirements.
- Many snap-fit problems can be traced to improper constraint.
- While constraint features are necessary and sufficient for a snap-fit attachment, enhancements are required to make the attachment robust and "world-class".
- Use generic descriptions of part shapes to transfer important snap-fit knowledge, past experience and lessons-learned between applications. Regardless of the application, the fundamental design principles for a specific basic shape combination will always be true.

2.4.2   Important Design Rules Introduced in Chapter 2

- In a fixed application, no relative motion between the parts is intended. The attachment is properly constrained when the mating part is constrained to the base part in exactly 12 DOM. In a moveable application, the attachment may be properly constrained in less than 12 DOM.
- An important rule for understanding and fixing snap-fit problems is that feature level problems cannot be fixed until one verifies there are no attachment level problems in the snap-fit.
- Select a mating part to base part engage direction so that the (opposite) separation direction is not in the same direction as any significant forces on the attachment.
- To ensure compatibility:
  All physical features in the interface must be compatible with the assembly motion.
  The selected assembly motion must be compatible with the basic shapes.
  The assembly and disassembly motions should be the same (although opposite in direction).
  Clearance must be provided for feature movements during assembly and disassembly.

3 Constraint Features

Constraint features are the locking and locating features that actually hold the parts together.

3.1 Introduction

Constraint is the most fundamental of the key requirements for a snap-fit and the features that provide positioning and strength in the attachment are the most important parts of a snap-fit. Many of the attachment level design rules that will be discussed in this and in later chapters involve getting proper constraint.

Recall the definition of a snap-fit from Chapter 1:

> A snap-fit is a mechanical joining system in which attachment occurs using *locating* and *locking features* (constraint features) that are homogenous with one or the other of the parent components being joined. Joining requires a flexible locking feature to move aside for engagement with the mating part followed by return of the locking feature toward its original position to accomplish the interference required to latch the components together. Locators, the second type of constraint feature, are inflexible, providing strength and stability in the attachment. Enhancements complete the snap-fit system, adding robustness and user-friendliness to the attachment.

Constraint features fall into two major groups: locators and locks. They were introduced in Chapter 2 and described briefly. This chapter will describe the details of constraint features and the concepts behind their use. Analysis of lock constraint features is discussed in Chapter 6.

In most of the illustrations in this book, radii at corners are not shown because of the complexity they add to the graphics creation process. However, the reader must keep in mind that a basic rule of plastic part design is to avoid sharp corners. This rule applies to both interior and exterior corners and to all features, including snap-fit features. Radii must be specified where the feature meets the parent material as well as at all the angles within the feature itself. The radius provides benefits to melt-flow for manufacturing, improves the resulting part quality and reduces stress-concentrations effects in loaded areas.

3.2 Locator Features

Discussion of constraint features begins with locators for two reasons: Locators are the *first constraint features considered* when we begin developing the snap-fit interface; second, they are relatively simple features compared to the considerably more complex and varied lock features locks.

By definition, locators are strong features. They provide part-to-part positioning (locating) and should also carry all significant forces in the attachment. During snap-fit development, locators are added to the interface in two ways. They are identified as pre-existing part features, like edges and surfaces, that can serve a locating function. These are called *natural locators*. Locators are also distinct features added to the interface specifically to do the locating function. Each has advantages. Natural locators are in the part already and do not add cost. Unlike distinct locator features, however, they are limited in constraint capability and do not, as a rule, provide for easy dimensional control or fine-tuning.

3.2.1 Locator Styles

Locators, being strong and inflexible, normally have no assembly deflection issues associated with them. Thus, unlike locks, they are relatively easy to understand and use. With few exceptions locators, if they are analyzed at all, usually require only a simple analysis of behavior under shear or compression loading. One notable exception is when a locator is used as a low-deflection lock feature. This is discussed in the lock features section.

The following definitions may seem unnecessary at first. Individual locators are identified in this manner in order to define them by name and characteristic shape. Locators that seem similar can have important performance differences. These will become apparent when one locator is paired with another in a locator pair. They will exhibit differences in degrees of motion removed, tolerance to dimensional variation and assembly motions allowed.

While there are infinite varieties of locators, we can organize most of them into three logical groups: *protrusion-like*, *surface-like* and *void-like*. In addition, because of the role they play when present in a snap-fit interface, living hinges are considered locators.

The first group consists of locators formed as a protrusion from a part, Fig. 3.1. Because they are protrusions, these locators are generally not natural locators.

3.2.1.1 Lug

Lugs are protruding locator features characterized by an "L" section intended to engage over an edge, Fig. 3.1a. Lugs are one of the most common locators and there are numerous variations on the basic "L" shape. One useful modification of the common lug is a *track*. A track is formed when two lugs face toward or away from each other and are extended to create a strong locator that allows for a slide assembly motion.

3.2.1.2 Tab

Tabs are flat protrusions with parallel or slightly tapered sides, Fig. 3.1b. They normally engage an edge or a slot.

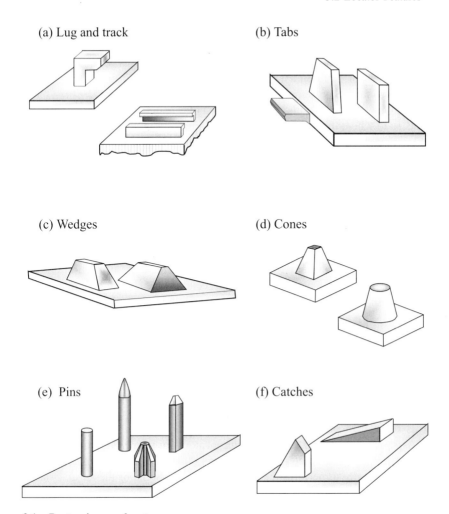

Figure 3.1 Protrusions as locators

3.2.1.3 Wedge

Wedges are a variation of the tab in which the base is much greater in area than a section towards the end, Fig. 3.1c. The greater thickness at the base makes them potentially much stronger than a tab. Wedges are intended to engage a slot and, like a cone, they provide constraint along the axis of the taper as well as in lateral directions. Wedges, by definition, have a base with a long and short axis.

3.2.1.4 Cone

Cones are a variation of the pin locator in which a section at the base is significantly larger than a section towards the end of the feature, Fig. 3.1d. Cones engage holes and, like the wedge, are intended to provide locating in the axial direction as well as in lateral directions. Cones may have a round or a square section. The square section cones look like wedges, but they are not. As a rule, because performance is identical and cones with a round section tend to be more robust for dimensional performance as well as easier to create in the mold, the round section cones are preferred over the square section cones.

Note that any features with thick sections, like cones and wedges, have important limitations as to where they can be placed on an injection molded part. Good mold design practice requires that thick sections be cored out, so appearance of the opposite surface and accessibility can limit use of these features.

3.2.1.5 Pin

Pins are features having either constant section or slight taper along the axis of symmetry, Fig. 3.1e. They may have round, square or complex sections. Pins generally engage holes, slots or edges and constrain only in lateral directions. Note that, in injection molded parts, features like pins can only have a truly constant section if they are formed in a plane parallel to the split line of the mold. Otherwise, draft angle requirements will mean that a slight taper exists.

3.2.1.6 Catch

Catches are *wedge-shaped* features, Fig. 3.1f. However, unlike the wedge feature, they are intended to engage against an edge, not into a hole or slot.

Let us summarize protrusion type locators by noting their roots in one fundamental feature. As shown in Fig. 3.2, all protruding locators are simply variations of a pin. Note again that none of the protruding locators are natural locators. In other words, they are added to the part specifically to perform the locating function. Some of the *surface-like* locators, discussed next and shown in Fig. 3.3, can be natural locators.

3.2.1.7 Surface

Surfaces are locally flat or smooth areas, Fig. 3.3a. They constrain in only one direction and are almost always natural locators.

3.2.1.8 Land

A land is a raised area on a surface, Fig. 3.3b. The land also provides a locating surface, but it allows local dimensional control and a fine-tuning capability that the (natural locator) surface may not be able to provide, at least not economically. Lands also permit clearance in some applications for ease of assembly, find more on that topic in the chapter about enhancements.

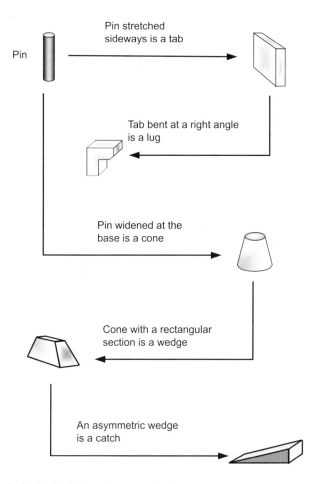

Figure 3.2 The pin is the basis for all protrusion locators

3.2.1.9 Edge

Edges are relatively thin areas; they are usually linear and orthogonal to a surface, Fig. 3.3c. An edge is generally on a part wall or on a rib or gusset and it can be either a natural or distinct locator, depending on the situation. Edges lend themselves quite readily to local dimensional control and fine-tuning.

Similarly to the protrusion locators and the basic pin, we can show how the surface-like locators can be evolved from a basic edge, Fig. 3.4.

The last group of locators is created by forming a void in a part, usually in a wall. Whereas the surface-like locators may be added or natural locators; voids, like protrusions, are almost always added specifically to perform a locating function.

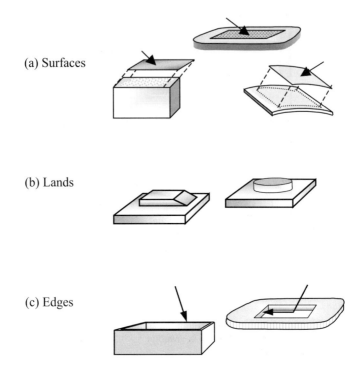

(a) Surfaces

(b) Lands

(c) Edges

Figure 3.3 Surface-like locators

3.2.1.10 Hole

Holes are openings in a panel or a surface. They may be round, square or some other shape, Fig. 3.5a. Holes, by definition, constrain in at least four degrees of motion. They may constrain in five DOM depending on the mating locator.

3.2.1.11 Slot

A slot is a hole elongated along one axis, Fig. 3.5b. The elongation serves to remove contact (constraint capability) along the long axis of the slot. By definition, a slot constrains in at least two and possibly three degrees of motion. Sometimes, the difference between a hole and slot is determined only by the nature of the mating locator. This is explained in the discussion of locator pairs.

3.2.1.12 Cutout

Cutouts are a hybrid of the hole and edge locators. A cutout has three active or useful edges rather than one, Fig. 3.5c. Like a hole, the cutout provides additional constraint capability; like an edge, it provides more assembly options. A cutout may *look* somewhat like a hole or a slot. Once again, the classification depends on how it is used.

3.2 Locator Features   53

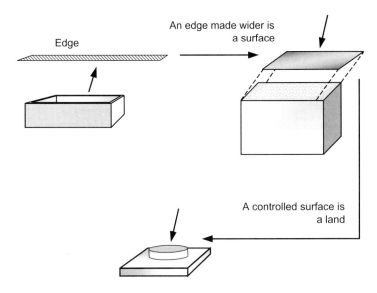

Figure 3.4   Edge as the basis for surface-like locators

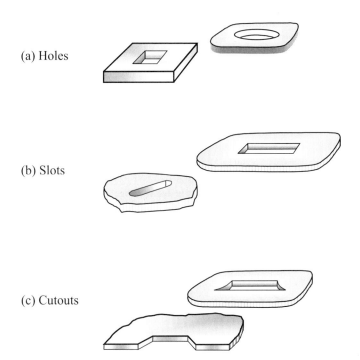

Figure 3.5   Voids as locators

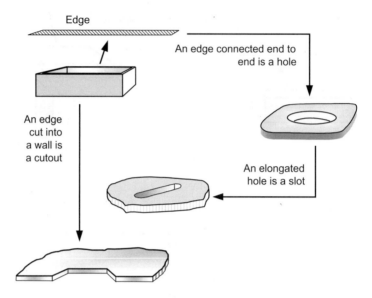

Figure 3.6 Edge as the basis for void-like locators

Figure 3.6 shows how the void-like locators also derive from the edge. Holes and slots are really an edge closed around on itself. Cutouts may be a fully closed edge or a three-sided edge configuration.

3.2.1.13 Living Hinge

A living hinge is a relatively thin connective section between two parts, Fig. 3.7. It joins the parts but also allows for (rotational) movement of one part relative to the other. In this sense it behaves like a locator pair in a moveable application. Because living hinges act as the first engaged locator pair and provide positioning as well as strength, they are classified as a locator rather than a lock.

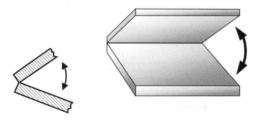

Figure 3.7 In a snap-fit, living hinges act as locators

3.2.2 Design Practices for Locator Pairs

This section describes how locators work together in pairs to produce effective constraint. Locators by themselves cannot provide constraint in the interface; they must be used in pairs. Thus we develop a snap-fit attachment using locator pairs, where a locator on the mating part engages a locator on the base part. Locator features and common locator pairs (except living hinges) are summarized in Table 3.1. Criteria for classifying some locator combinations as "N—Possible but not recommended" in this table include:

- Would create internal stress in the attachment between locator pairs.
- Some natural locators against natural locators can be difficult to fine-tune.
- Some adjustable locators against other adjustable locators would be redundant.
- Some combinations are inherently weaker than a preferred alternative.

Figure 3.8 shows some examples of how identical or similar features can have different constraint and assembly characteristics, and different names, depending on the other locator in the pair. Figure 3.8a shows how a pin-hole locator pair differs from a cone-hole pair in degrees of motion removed. Figure 3.8b shows how a wedge-hole differs from a wedge-slot pair. Figure 3.8c shows how a rectangular opening provides an edge in a lug-edge and how that simple edge differs from a cutout in the lug-cutout pair. The same rectangular opening in Fig. 3.8b and 3.8c is used as three different kinds of locator. In other words, a locator feature is not fully defined until both members of the locator pair are identified.

Note how the constraint characteristics of the protrusion-like locators are defined independently of the surface or edge to which they are attached.

3.2.2.1 Terminology

We have just seen in Fig. 3.8 how locator pairs provide constraint in specific directions. At this point, it is appropriate to define some terms. In mechanics, we can think of a force as acting along a "line-of-action", Fig. 3.9a. Constraint consists of two abilities. One is the ability to provide positioning. For the catch-edge locator pair, Fig. 3.9b, we can show positioning with an arrow (a directional vector) representing resistance to movement, Fig. 3.9c. The second is the ability to react against potential forces. This is strength and can be shown as an arrow in the direction of resistance to a force, Fig. 3.9d. A frame of reference for identifying position and strength direction is needed. In all the illustrations in this book, unless noted otherwise, constraint capabilities are noted in terms of their effect on the mating part. Where no mating part is shown, as in these figures, an "R" identifies the reference feature.

It is no surprise that position and strength capabilities occur in the same directions along the same line of action, Fig. 3.9e. However, there are times we wish to differentiate between position and strength capabilities of a locator so we may refer to positional lines of action or strength lines of action. Keep in mind they are co-linear although they may have different levels of performance, i.e. high strength but low positional accuracy, high strength and precise positional accuracy, etc. In discussing interactions between locator pairs, we can compare the appropriate lines of action, Fig. 3.10.

Table 3.1 Locator Feature and Locator Pair Summary

		Locator name and *maximum* possible DOM removed											
		Protrusion-like						Surface-like			Void-like		
		Lug 4	Tab 3	Wedge 5	Cone 5	Pin 4	Catch 3	Surface 1	Land 1	Edge 1	Hole 5	Slot 3	Cut-out 4
Protrusion-like	Lug 2	Track*								C			C
	Tab 1	Track*								C			N
	Wedge 3							N		●		C	N
	Cone 5										C	●	N
	Pin 2					N○			N○	N	C	N○	N
	Catch 1					C	C	C	N○	C	C	●	●
Surface-like	Surface 1		N○	●	N		C	●	C	●	C	C	R
	Land 1						N○	N○	●	C			
	Edge 1	C	C			C	C	C	C	C			R
Void-like	Hole 4									R			
	Slot 2												
	Cut-out 3	C	N○					R ●	N○	R			R

*—Special case.
Empty cell—Not possible given the locator definition.
C—Common design situation.

R—Rare but possible.
N—Possible but not recommended. Use indicated pair (●) instead.
Preference is based on general strength considerations.

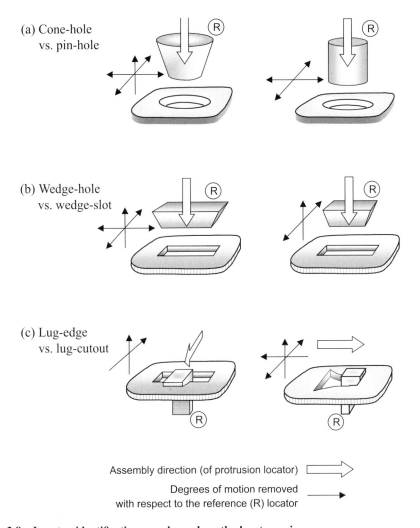

Figure 3.8 Locator identification can depend on the locator pair

3.2.2.2 Locator Pairs, Constraint and Strength

In general, because locators are strong relative to locks, the more degrees of motion that can be removed with locator pairs in a snap-fit, the stronger the attachment. This is an extremely important point and means, in essence, the more locators and fewer locks, the better. The obvious follow-up question is "How can I get more locators and fewer locks into a snap-fit to increase its strength?" The answer begins with consideration of the spatial element *assembly* motion, which was introduced in Chapter 2. An extremely important principle of snap-fits is: *Snap-fit attachment strength is, first of all, determined by the assembly motion selected for the application, not by the locking feature strength.*

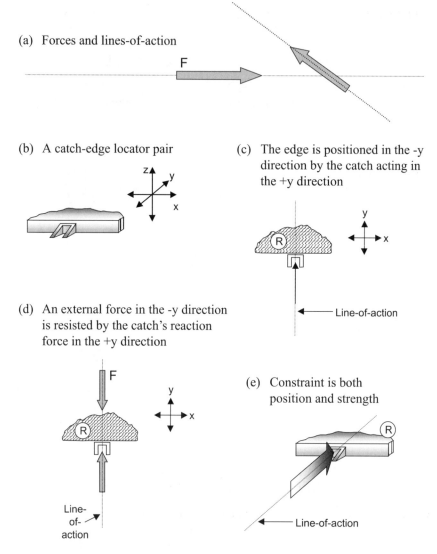

Figure 3.9 Terminology

Locator pair selection for a given application is a function of the assembly motion. We can identify the ability of each of the five generic assembly motions to maximize the DOM removed by locators and minimize DOM removed by locks.

As a general rule, locator pairs that remove the most degrees of motion (DOM) are desirable from a design efficiency standpoint. But combining too many of these in one interface will cause over-constraint. Another desirable characteristic of a locator pair is that it help resist any possible forces in the separation direction. The only locator pairs with this

3.2 Locator Features 59

(a) Example: catches acting against edges

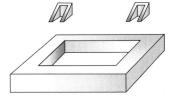

(b) Catches are co-linear and of opposite sense

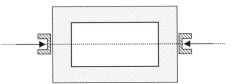

(c) Catches are co-linear and of the same sense

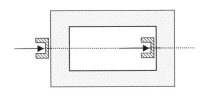

(d) Catches have the same sense and parallel lines-of-action

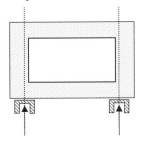

(e) Catches of opposite sense with parallel lines-of-action

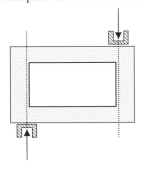

(f) Catch lines-of-action are perpendicular

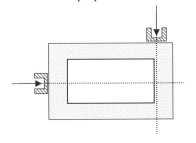

Figure 3.10 Lines-of-action

ability are the lug-edge, lug-cutout and living hinge. There are other high-strength snap-fit arrangements, but these three are unique in their use of a locator feature to help carry load in the removal direction. The only assembly motion that supports this ability is the tip, Fig. 3.11.

Table 3.2 shows how the (recommended) locator pairs from Table 3.1 are related to degrees of motion removed and assembly motion.

When considering the overall effect of assembly motion on locator and lock pair selection, the potential for degrees of motion removed by locators is highest with the slide, twist and pivot motions, next highest for a tip motion and lowest for the push assembly motion, Table 3.3. This is strictly a function of how all of the locator pairs can be arranged in the interface to accommodate the assembly motion. Note that while the slide, twist and pivot motions may remove the highest DOM in some scenarios, the tip motion is preferred from a total optimization standpoint. As a general rule, the push assembly motion, probably the

(a) A push motion requires that separation forces be resisted by locks

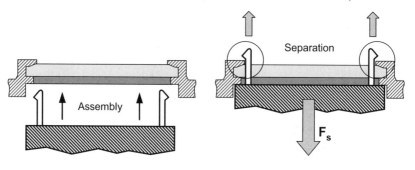

(b) A tip motion allows some separation forces to be resisted by locators

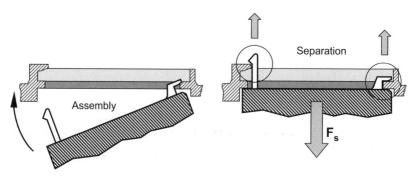

Figure 3.11 Assembly motion and constraint feature selection

3.2 Locator Features   61

Table 3.2 Locator Pairs, Degrees of Motion, and Assembly Motion Summary

	Notes	Locator pair	DOM removed	Assembly motion				
				Push	Slide	Tip	Twist	Pivot
Possible locator pairs	* p-s	Track	10		✓			
	*	Living Hinge	10			✓		
	v-v	Cutout-Cutout	5	✓		✓		
	p-v	Cone-Hole	5	✓		✓		
	p-v	Lug-Cutout	4			✓		✓
	p-v	Pin-Hole	4	✓		✓		
	p-v	Wedge-Slot	3	✓		✓		
	p-v	Catch-Cutout	3	✓		✓		
	s-v	Surface-Cutout	3	✓		✓		
	s-v	Edge-Cutout	3	✓		✓		
	p-v	Pin-Slot	2	✓		✓		
	p-s	Lug-Edge	2		✓	✓	✓	✓
	p-v	Tab-Slot	2	✓		✓		✓
	p-s	Tab-Edge	1	✓	✓		✓✓	✓
	p-p	Catch-Catch	1	✓	✓	✓	✓	
	p-s	Catch-Surface	1	✓	✓	✓	✓✓	✓
	p-s	Catch-Edge	1	✓✓	✓✓	✓	✓✓	✓
	s-s	Surface-Land	1	✓✓	✓✓	✓	✓	✓
	s-s	Surface-Edge	1	✓✓	✓✓	✓	✓	✓
	s-s	Land-Edge	1	✓✓	✓✓	✓	✓	✓
	s-s	Edge-Edge	1	✓	✓✓	✓		

* Special case
s-s surface-surface
s-v surface-void
p-s protrusion-surface
p-v protrusion-void
v-v void-void
p-p protrusion-protrusion
Bold (✓) indicates possible use as first engaged pair for the assembly motion.
Light (✓) indicates use as second or third engaged pair for the assembly motion.

Table 3.3 Assembly Motion and Degrees of Motion

Assembly Motion	Best case scenario		Worst case scenario		Ease of Use
	Maximum possible DOM removed by all locators	Remaining DOM to be removed by locks	Minimum possible DOM removed by all locators	Remaining DOM to be removed by locks	
Slide	11	1	10	2	Limited by basic shapes
Twist	11	1	10	2	Limited by basic shapes
Pivot	11	1	10	2	Limited by basic shapes
Tip	10	2	10	2	High adaptability
Push	7	5	7	5	High adaptability
DOM TOTALS	12 DOM		12 DOM		

most frequently used motion, should be avoided whenever possible because it requires the most degrees of motion removed by the lock features.

3.2.2.3 Locator Pairs and Ease of Assembly

The part assembly motion will require selection and orientation of locator pair combinations that permit assembly. A typical compatibility problem is a situation in which the designer fails to recognize that the locator pairs in the attachment will not allow the assembly motion to occur as anticipated. This situation was discussed in Chapter 2 and illustrated in Fig. 2.7. Thinking about the application in terms of assembly motion has a number of advantages. In addition to encouraging constraint feature decisions leading to maximum strength, it increases design creativity (see Chapter 7) and designing with the available assembly motions in mind will eliminate the possibility of motion/access/constraint feature compatibility.

The first locators considered during design should be the one(s) that make first contact during assembly. This locator pair(s) should also provide the guidance function (a required enhancement) as shown in Fig. 3.12. Another advantage of the lug feature is shown here; lugs can serve as assembly guides as in Fig. 3.12b. No additional guide enhancements, like the pins in Fig. 3.12a, are needed. If a pilot is necessary, the first locator pair can also provide that function. Pilots ensure that the mating part can only be assembled to the mating part in the correct orientation. These and other enhancements are discussed in detail in Chapter 4. Once these first locator pairs are in place, the remaining locators can be added.

3.2.2.4 Locator Pairs and Dimensional Control

Some pairs in an application may be identified as "position-critical" because they will control important positioning or alignment behavior of the parts. For this reason, they will be potential sites for fine-tuning the attachment. Keep this in mind as these sites are identified and use caution if two natural locators make up a position-critical pair where fine-tuning

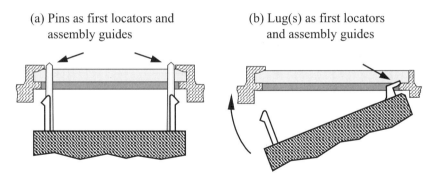

Figure 3.12 First locators to make contact during assembly should also serve as guides

may be necessary. Making in-mold changes to major part features like natural locators can be difficult and costly. Edge-surface (natural locator) pairs can be changed to land-edge pairs as shown in the solid-to-opening application in Fig. 3.13. If flushness is also important in this application, fine-tuning at the surface-surface natural locator that controls positioning along the z-axis would be required. Lands could be added to one of those surfaces (the ledge around the opening, for instance) to support easy fine-tuning for flushness. The subject of designing for feature fine-tuning is covered in more detail in the enhancements chapter.

Locator pairs acting together to constrain the same rotation or translational movement should be placed as far apart from each other as possible to maximize part stability and minimize sensitivity to dimensional variation; see Fig. 3.14. In this solid to surface example, we will only be concerned with constraint in the x–y plane.

The effect of locator pair spacing on dimensional stability is a simple inverse relationship. In Fig. 3.14c, the parallel lines-of-action of the locator pairs catch #1 and catch #2 (C-1 and C-2) are distance (d) apart. The resulting effect on point a's position along the x-axis is a function of the ratio h/d so that the tolerance of C-1 to C-2 in the y direction is:

$$\Delta a = \frac{h}{d} \tag{3.1}$$

If the y-tolerance of C-1 to C-2 is ±0.1 mm and $h/d = 2.5$, then the effect on point a's position is ±0.25 mm; calculated from $\Delta a = 2.5(\pm 0.1 \text{ mm})$.

In Fig. 3.14d, the parallel lines-of-action of locator pairs (C-1 and C-2) are much farther apart. If the ratio of h/d is 0.67, the effect on point a's position along the x-axis is now:

$$\Delta a = \frac{h}{d} \delta a = 0.67[0.1 \text{ mm}] = 0.067 \text{ mm} \tag{3.2}$$

Point a moves in the x-axis, therefore any other locator pairs with constraint in the x-axis will be directly affected by the relation between catches #1 and #2. The lug#1-edge constraint pair will be affected and, to a lesser extent, so will the lug#2-edge constraint pair.

Note that there will be other factor affecting the tolerance on point a's position. The effect of the locator pair positions should be included in the dimensional evaluation of the part.

The effects of locator pairs acting as couples are similar to those described above for translational motion. Couples are illustrated in Fig. 3.14 e and f. In mechanics, a couple is

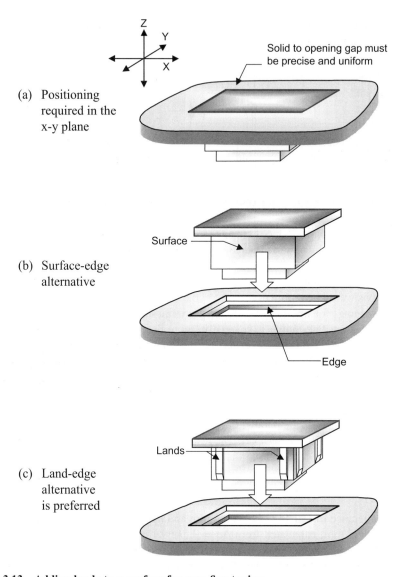

Figure 3.13 Adding lands to a surface for easy fine-tuning

defined as two equal forces, of opposite sense, having parallel lines-of-action. Couples act to produce a pure rotational force or prevent rotational motion.

When possible, the position-critical locator pairs should be used as the datum for dimensioning all the other constraint pairs in the interface. The datum for the position-critical locators should be the related alignment site.

To minimize the effects of mold tolerances and plastic shrinkage, position-critical locator pairs should be placed as close as possible to the site where alignment is required.

3.2 Locator Features 65

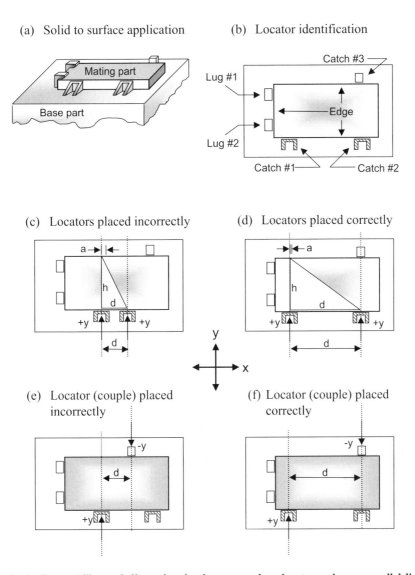

Figure 3.14 For stability and dimensional robustness, place locator pairs so parallel lines-of-action are as far apart as possible

Some applications may not need critical positioning or alignment and this requirement can be relaxed. Note too that the position-critical locators are not necessarily the first locator pairs engaged during assembly.

3.2.2.5 Locator Pairs and Mechanical Advantage

Unlike dimensional stability, translational strength is not affected by the distance between locator pairs having parallel lines-of-action and the same sense, Fig. 3.15c. (It is possible to

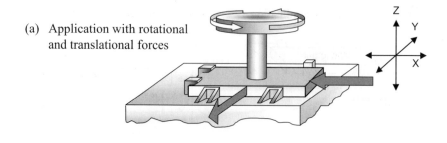

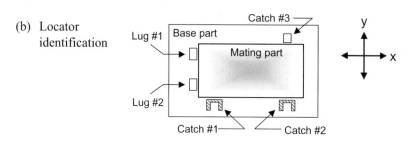

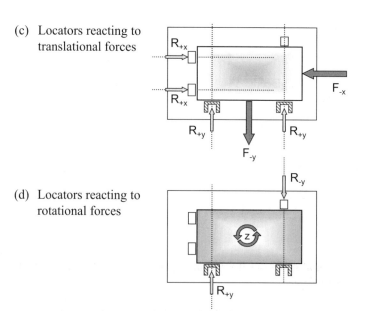

Figure 3.15 For strength, place locator pairs acting as couples as far apart as possible

remove translational movement along an axis with one locator pair.) For removing degrees of motion in rotation, however, two locator pairs must work together as a couple and, as with positioning, distance does have an effect. For maximum mechanical advantage against rotational forces, locator pairs acting together as a couple should be placed with their (parallel) lines-of-action as far apart from each other as possible, Fig. 3.15d.

3.2.2.6 Locator Pairs and Compliance

Compliance is treated as an enhancement and is discussed in detail in Chapter 4. For our purposes here, it is sufficient to define compliance as tolerance to dimensional variation. In other words, designing compliance into the snap-fit interface allows us to use normal tolerances and maintain a close, rattle-free fit between parts.

Compliance can have an effect on a locator pair's strength or positioning capability. Caution is required if compliance must be designed into position-critical or load-carrying locator pairs.

3.2.2.7 Locators Summary

Locators are strong and inflexible constraint features. Their job in a snap-fit is to provide both positioning of the mating to the base part and strength to prevent motion under external forces.

This section introduced locators as individual features and then explained how they operate in a snap-fit as locator pairs. Functional issues and design rules associated with locator pairs in an application were then explained.

3.3 Lock Features

Locks are the other constraint features and they have traditionally represented snap-fit technology. To many designers, lock features, particularly cantilever hooks, *are* snap-fits. However, at the attachment level, locks are just one part of the system. Locks hold the mating part to the base part so the strong locating features can do their job. Along with locators, locks are the "necessary and sufficient" features for a snap-fit attachment. From the snap-fit definition:

> Locks are relatively flexible features. They move aside for engagement then return toward their original position to produce the interference required to latch parts together.

The fundamental problem in snap-fit design is that locks must be weak in order to deflect for assembly yet strong enough to prevent part separation. Sometimes locks should release but only under certain conditions. These complex and sometimes conflicting requirements, plus the need for analysis of deflection and strength, make lock features much more difficult to design than locators.

3.3.1 Lock Feature Styles

Locks are identified and grouped by their fundamental differences in assembly and retention behavior. These differences are most obvious when we consider the calculations necessary for evaluating assembly and retention behavior. Most locks require some form of *flexible* behavior to permit assembly. Retention behavior is expected to be rigid until or unless disassembly is desired. It is logical to infer varying degrees of retention performance depending on the nature of the retention behavior and, indeed, we find that the effectiveness of bending is far less than tensile, compression or shear behavior in resisting lock release.

Lock styles are defined here in a general order of usage. The cantilever beam style is by far the most commonly used lock feature. Trap and planar locks are also relatively common. Annular and torsional locks are less common. Brief definitions of the five styles of flexible lock are given here then are followed by detailed discussion of each.

- *Cantilever beam* locks engage through beam bending and retain through the mechanics of beam tension and bending or beam tension and shear.
- *Planar locks* involve one or two deflecting walls, usually with an edge and a catch on the walls. They engage through plate deflection and retain through shear or compression strength and plate mechanics.
- *Trap* locks engage through beam bending (like the cantilever beam locks) but they retain through beam compression. This is a significant difference and we will see that traps can be extremely strong locks.
- *Torsional* locks use torsional behavior for assembly deflection. Retention also depends on the nature of the torsion member.
- *Annular* locks use interference between concentric ridges on the internal and/or external walls of cylinders and rely on radial elasticity for assembly and retention strength.

The assembly and retention behaviors are much different for each of these lock styles. This makes each style better suited for some applications and worse for others. Keep in mind, however, that for many applications, the lock style must be selected to satisfy other requirements such as available space for lock deflection, die movement for part manufacturing and access for assembly. The most feasible lock style may not always be the best lock style. But the versatility and variety of lock features allows many options and solutions to a design situation.

As with locators, note that the parent material, such as a surface, on which a lock is mounted is *not* considered part of the lock feature. If the parent material provides constraint, it is considered a locator feature. For clarity and illustrative purposes, it is usually convenient to show a lock feature with a surface, just remember it is *not* part of the lock.

3.3.2 Cantilever Beam Locks

The cantilever beam style lock is by far the most common locking feature and it exists in infinite variety. Because it is the most common lock style we will spend much more time on it than on the others. For the same reason, when appropriate to do so, it is used throughout

The deflection mechanism is a beam
The retention mechanism is a catch

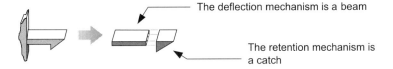

Figure 3.16 Major parts of a lock feature

the book when a lock is needed to complete an example or an illustration. Many principles of lock behavior, particularly those associated with the retention mechanism, introduced in the cantilever beam section will also apply to the other lock styles.

All locks have two major components, a *deflection mechanism* that allows for assembly and separation and a *retention mechanism* where contact occurs with the mating constraint feature, Fig. 3.16. It is helpful to consider them separately.

3.3.2.1 The Deflection Mechanism

In a cantilever lock, the deflection mechanism is a beam and there are as many kinds of deflection mechanisms, as there are possible beam shapes and sections. Some of the more common beam shapes are shown in Fig. 3.17a. The beam can also vary by section and some of the more common beam sections are shown in Fig. 3.17b. Analysis of beam behavior for assembly is based on the classical bending equations for a cantilever beam fixed at one end. Analysis for retention depends the retention mechanism style.

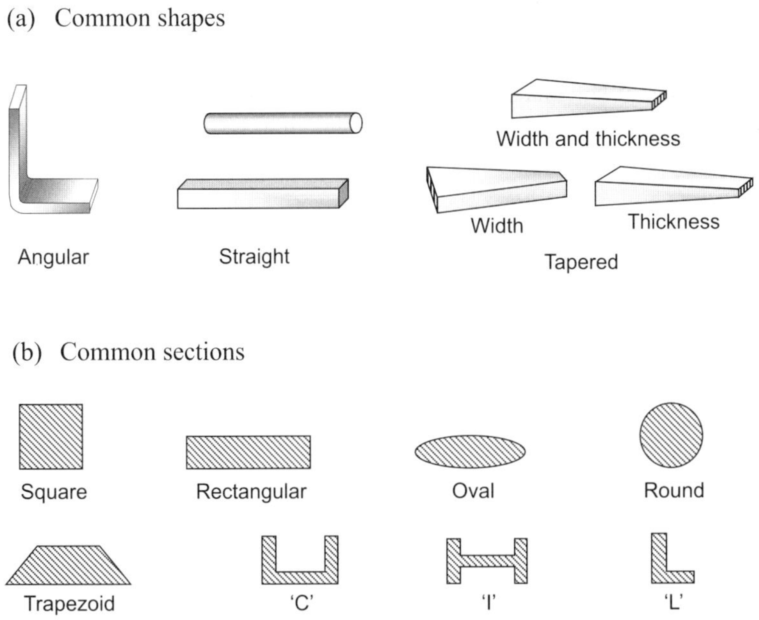

(a) Common shapes
Width and thickness
Width
Thickness
Angular
Straight
Tapered
(b) Common sections
Square
Rectangular
Oval
Round
Trapezoid
'C'
'I'
'L'

Figure 3.17 Common beam shapes and sections

By far, the most common lock configurations use beams similar to those shown in Fig. 17a with a rectangular section. The other sections are possibilities and may be useful in solving a unique problem, but they are not generally recommended because they can make analysis more difficult and/or add complexity to the mold.

3.3.2.2 The Retention Mechanism

The retention mechanism on the beam can be selected independently of the beam itself. This increases the cantilever lock design options because beam and retention mechanism styles can be mixed to suit the application. The most common retention mechanism is some form of protrusion style locator, as shown in Fig. 3.18. When the retention mechanism is a protrusion, the cantilever lock is called a hook because it "hooks" over an edge to engage. The hook style cantilever locks must, of necessity, resist separation through beam bending.

The inherent weakness of the hook lock is that when separation force is applied to the lock, the reaction force cannot be along the neutral axis of the beam, regardless of the shape of the mechanism. As shown in Fig. 3.19, there will always be an offset (d). Thus the hook is destined to bend. Unfortunately bending is the cantilever beam's weakest direction for resistance to deflection, Fig. 3.19a.

Even non-releasing hooks with a retention face angle at or near 90° can release under a sufficiently high force. When a non-releasing hook does release under load, the typical pattern of release begins with initial distortion of the beam at the retention mechanism. This causes a reduction in the retention face angle which then enables additional slippage along the retention face and beam bending for release, Fig. 3.19b. The kinds of hook failures shown in Fig. 3.19c are very unlikely unless the hook end is restrained from rotating.

When an angle greater than 90° is used on both the hook and the mating feature, then high strengths are possible, Fig. 3.19d. This kind of lock application is frequently found on the closure buckles of soft-sided hand baggage like children's backpacks, book bags and laptop computer cases and they are quite strong. It does require enough clearance or

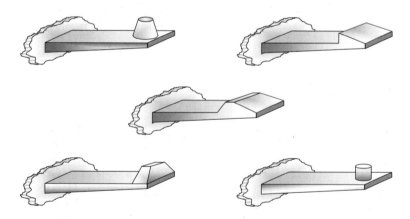

Figure 3.18 Common retention mechanisms based on protrusion-like locators

3.3 Lock Features   71

(a) Bending resistance to separation

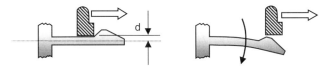

(b) A non-releasing hook will not prevent separation

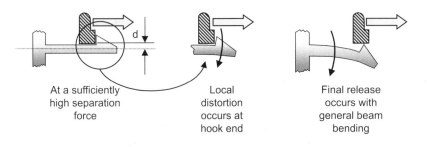

At a sufficiently high separation force

Local distortion occurs at hook end

Final release occurs with general beam bending

(c) Highly unlikely failure modes

Shear failure under the catch

Tensile failure of the beam

(d) A retention face with a reverse angle can prevent separation in a non-releasing hook.

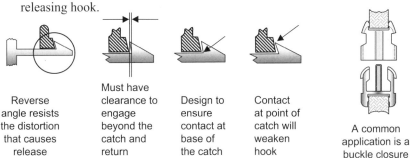

Reverse angle resists the distortion that causes release

Must have clearance to engage beyond the catch and return

Design to ensure contact at base of the catch

Contact at point of catch will weaken hook

A common application is a buckle closure

Figure 3.19   The inherent weakness of a hook-style cantilever lock

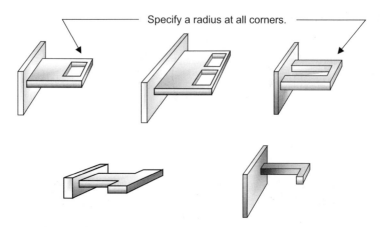

Figure 3.20 The loop-style cantilever lock

compliance in the system to allow the lock face to move past the engagement point and then return. For this reason, it is not practical in many applications.

A cantilever lock that uses void or edge-like retention mechanisms at the end of the beam is inherently stronger than the beam/catch hook, Fig. 3.20. This lock style is called a "loop" [1] because it somewhat resembles a loop of rope thrown over a post. Just like a rope, which has no strength in bending but is extremely strong in tension, the loop version of the cantilever lock can have extremely high retention strength because it relies on tensile, not bending, strength for retention. The "T" and "L" configurations are simply variations of the basic loop. A "T" is a loop split down the middle and reconnected along the outside edges; an "L" is one-half of a loop.

When a loop is used as a non-releasing lock, the reaction force is in line with the neutral axis of the beam and no bending can occur. Instead, retention strength is determined by the loop's dimensions and the tensile and shear strengths of the materials that make up the lock pair, Fig. 3.21. This characteristic means the loop style lock can always provide better

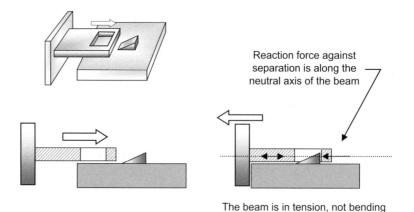

Figure 3.21 Retention strength advantage of the loop-style cantilever lock

retention in a given application than a hook. Often the mating feature to a loop hook in the lock pair is a catch, which is also an inherently strong feature. The loop hook and catch together can be an extremely strong lock pair and, unlike a hook lock, tends to be resistant to release under shock loading.

In the case of releasing locks, the reaction force is no longer along the neutral-axis of the loop, but the loop still has retention advantages. In addition to advantages in retention, the loop enjoys other advantages over the hook. It inherently has a more desirable force-deflection signature for assembly as discussed in the following sections. These additional retention and assembly characteristics will be discussed shortly.

The loop also requires less clearance for deflection and can deliver equivalent or better retention performance when space for lock deflection is limited.

An issue unique to the loop is the likelihood of forming a knitline somewhere in the loop during the manufacturing process. Knitlines occur where two fronts of plastic material meet as the melt flows through the mold. The loop's shape practically guarantees this will happen, Fig. '3.22a. Knitlines may reduce the strength of the material at that point. Test data indicates the effect may be as much as a 65% strength reduction depending on the material and the absence or presence of a filler [2]. In addition to the material itself, the amount of strength reduction depends on the temperature (i.e. viscosity) of the fronts and the ability of the surface layers of the two fronts to merge. The strength reduction is most dramatic in filled materials; the fibers will not flow across it so the knitline consists only of the polymer

(a) Knitlines are almost guaranteed in loops

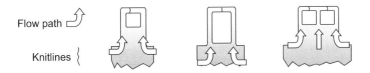

(b) Location makes a difference

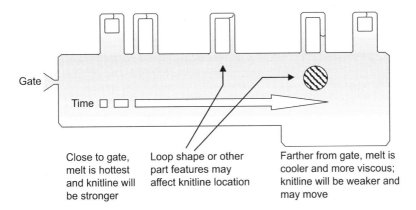

Figure 3.22 Knitlines in the loop-style cantilever lock

material, making it significantly weaker. In the tests cited above, unfilled polypropylene showed a 14% reduction and a 30% glass-filled polypropylene showed a 66% reduction in strength. The unfilled and 40% glass-filled nylon 66 test results were 3% and 48% respectively. (These results occurred under specific test conditions and should not be considered design data.)

Loops having identical shapes but located in different areas of the same part can have different levels of knitline strength and the knitlines can occur at different locations, Fig. 3.22b. This is due to local flow characteristics and because the melt temperature at a given point depends on its distance from the gate and cooling effects of the mold along the flow path.

Beall [2] recommends adding a drawing note indicating "No weldline (knitline) in this area" as a precaution for any highly loaded area of a part. This is a good idea, but the shape of the loop may make knitline prevention impossible. In general, the designer should accept that knit lines will occur in the loop and design to compensate for them; possible designs are shown in Fig. 3.23. Study and testing of prototype parts will indicate actual knitline location(s) and allow verification of the effectiveness of the solutions.

As mentioned elsewhere, it is good design practice to specify fillets and radii on all corners, both internal and external, on plastic injection-molded parts. This is especially critical in loops, where a sharp corner in the opening becomes a weak site due to molded-in stresses and also a stress riser under loads. They may not always be shown in the illustrations, but always specify a radius on all corners of a loop.

3.3.2.3 Cantilever Lock Examples

Some additional examples of cantilever locks are shown in Fig. 3.24. Although they are shown as extending at 90° from the plane of a wall, any angle is possible. Cantilever locks can also extend from an edge at any angle or they can be in-plane, extending from an edge or lying within the boundaries of a wall, Fig. 3.25.

3.3.2.4 Locators as Cantilever Locks

We recognize lock features by their deflection and, in general, we think of that deflection in terms of relatively large movements. However, locators can sometimes be used as locks

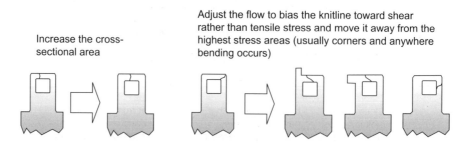

Figure 3.23 Compensating for knitline weakness

(a) A common hook, showing reasonably good proportions

(b) Extra long beam may warp and is too thin relative to length for good retention strength

(c) Beam is too short relative to thickness and insertion face is too steep

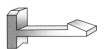

(d) Turning the beam back on itself gives extra length for deflection in confined spaces but may be difficult to mold

(e) Another version of a curved beam

(f) Any of these examples can also be a loop style lock

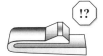

(g) Turning the retention feature 90° can improve performance significantly; see Chapter 5

(h) The basic loop

(i) Another example of a retention feature turned 90°

!? Locks with this mark are not recommended or require extra care

Figure 3.24 Variations of the cantilever lock

(j) Performance may be adjusted by adding a rib, but not where it will be under tensile stress

(k) Another example of ribs in tension

(l) If you must use them, put ribs where they will be in compression

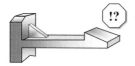

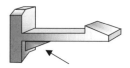

(m) Curved section, usually appears as one segment of a ring of these locks

(n) Bad design, all the strain is concentrated at the hook's base

(o) The beam and retention mechanism are the same

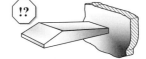
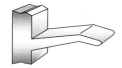

(p) Multiple locks at one site

(q) The beams in any of these locks can be tapered in thickness for improved strain distribution

(r) Tapering the beam on the width is also possible

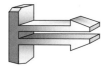
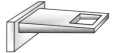

(s) Beam fixed at both ends

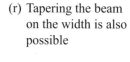 Locks with this mark are not recommended or require extra care

Figure 3.24 (*continued*)   **Variations of the cantilever lock**

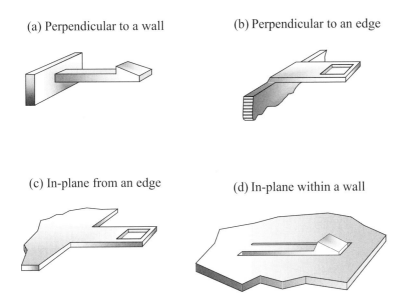

Figure 3.25 Lock orientation to parent material

when the required deflections are very low. The two requirements for these applications are an assembly motion that involves sliding (the slide, twist and pivot motions) and low or no force in the separation direction. Figure 3.26 shows some examples. In all these cases, the (lock) locator feature deflects over a small interference feature then returns to its relaxed state. The principles of constraint, beam deflection, tolerances and strength that apply to all lock pairs still apply.

3.3.2.5 Lock Pairs

As with locators, a lock feature on one part requires a mating feature on the other part. Together they make up a *lock pair*. In a lock pair, the mating feature is usually a locator feature such as an edge or catch. But there is no reason why, in some cases, it cannot be another lock. This can be useful when getting enough deflection out of only one feature is difficult. Lock pairs are important because lock effectiveness is a function of both members of the pair. We cannot isolate the deflecting lock feature and expect to understand or predict its performance.

The deflecting member of a lock pair may be placed on either the mating part or on the base part. Sometimes this is an economic/risk decision where it may be wise to put the lock on the smaller and less expensive part because the lock may be subject to damage during service/removal. Other times, lock placement is based on performance, putting the lock on the part with the best material properties to support desired lock performance.

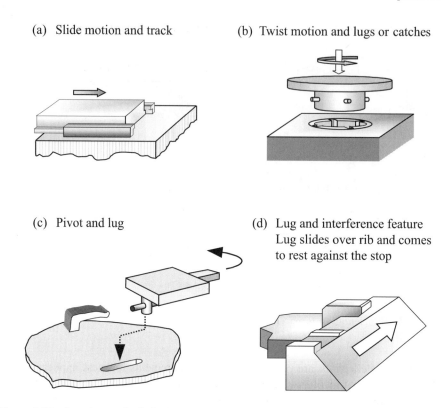

Figure 3.26 Locators as lock features

Unlike locator pairs, lock pairs will usually only remove one DOM. For cantilever hooks, traps and torsional locks expect lock strength only in the direction that resists separation. Annular and planar locks can sometimes provide strength in more than one DOM.

In the following discussions of lock behavior, the lock is treated as engaging a locator feature as the other member of the lock pair.

3.3.2.6 Cantilever Lock Assembly Behavior

With the common cantilever hook, the insertion face angle increases as the (hook) lock pair is engaged, Fig. 3.27. This causes the assembly force signature to increase geometrically, Fig. 3.28a. The resulting high final force can sometimes cause difficult assembly and feedback to the operator may be poor. If this is the case, a profile added to the insertion face, Fig. 3.29, can make the assembly signature more operator-friendly by reducing the maximum assembly force and changing the "feel" of the lock. Depending on the shape of the profile, the signature can be made to have a constant rate of change or a decreasing rate, Fig. 3.28b and c. The latter will give an "over-center" feel to the assembly and is most preferred for operator feedback.

3.3 Lock Features   79

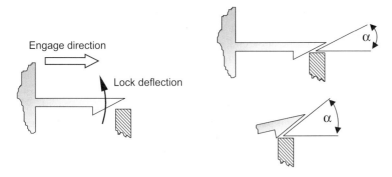

Figure 3.27 Increase in the insertion face angle during assembly; the most common situation

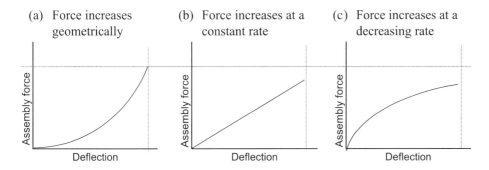

Figure 3.28 Typical assembly force-deflection signatures. Note that the maximum assembly force is lower at the same deflection in (b) and (c)

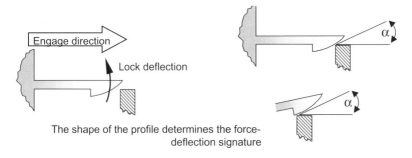

Figure 3.29 Adding a profile to the hook insertion face

The profile shape can be calculated to give the desired signature. A simplified calculation (see Chapter 6) can be based on treating the beam as if it is bending from its base with no curvature and no rotation of the hook end; often this is sufficient. More complex calculations accounting for beam curvature and end rotation can be applied if desired.

In an application where the action is controlled-moveable, the customer may be using the lock frequently. In this instance, the nature of the force-deflection signature can give the user the perception of either high quality or poor quality. The insertion face should always have a profile to improve user-feel and perceived quality and, again, an over-center signature as in Fig. 3.28c is preferred. High assembly forces repeated many times in a moveable application may also cause eventual damage to one or both members of the lock pair. An insertion face profile can reduce the chances of long-term damage by reducing the maximum assembly force.

Not all hooks exhibit this geometrically increasing assembly signature. If the nature of the hook is such that the insertion face angle does not change with the assembly movement, then the assembly signature will have a constant slope as shown by the loop style lock in Fig. 3.30. In this case too, the catch profile can be modified to give an over-center feel if desired.

One caution with respect to the loop design shown in Figs. 3.20a and b is that the walls at the open retention area are relatively weak compared to the remaining part of the beam. Design to ensure that assembly bending and strain is not concentrated in those walls. A solution is to extend the opening to the base of the beam as in Fig. 3.20c.

3.3.2.7 Cantilever Lock Retention and Disassembly Behavior

Some principles of retention were introduced with the discussion of the retention mechanism, but there are a few more that should be covered now. As with the assembly behavior, this is a qualitative discussion of cantilever hook behavior. Quantitative analysis of assembly and retention behavior is covered in Chapter 6.

As a general rule, it is recommended that lock features carry no significant forces in the separation direction. This is because locks tend to be relatively weak in that direction although, as we saw with the loop, some cantilever style locks can be quite strong. Design reality, however, is that many locks will, from time to time, be required to carry forces in the separation direction.

(a) The insertion face angle is on the catch, not the lock

(b) The angle stays constant, resulting in a force-deflection signature as shown in Fig. 3.28b

(c) Adding a curved profile to the catch results in a signature like Fig. 3.28c

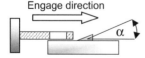

Figure 3.30 Assembly behavior of the loop-style lock

It is important to differentiate between the kinds of forces to which a lock might be subjected. Separation forces may be low, which is good, but if they are continuous and long-term they may result in plastic creep and lock release. Forces may be high but transient and, in a properly designed application, have no effect. In an application with poorly designed locks, the same transient forces may cause unintended separation. It is these transient forces we are concerned with here. Of course, the terms "high, low, long-term, short-term" are relative and depend entirely on the mechanical properties of the plastic(s) in any given application.

One of the most common mistakes made in snap-fit design is to use the lock to react against forces other than those in the separation direction. This results in an under-constraint condition because most locks, and certainly the cantilever hook, are intended to constrain in one direction only, the separation direction, Fig. 3.31. Always ensure that locator features are present to carry these other forces. Generally, the locators should be close to the lock for maximum effect.

What happens during application of a transient force to a lock? The energy is either absorbed by the locking system or the lock releases. The goal is to absorb the energy before the lock releases and without permanent damage to the lock.

Remember, even some locks designed to be non-releasing will release under sufficient force; refer back to Fig. 3.19b.

Figure 3.32a shows a typical cantilever hook. It has an angle less than 90° on the retention face, indicating it is probably a releasing lock. But, as with the lock in Fig. 3.19b, the mechanics of separation will be the same as for a non-releasing lock. As a separation force is applied, the hook begins to bend and the retention face angle decreases, Fig. 3.32b. (This is the opposite of the effect beam deflection had on the insertion face angle.) As the retention face angle decreases, its contribution to resisting the separation decreases. This decrease in the retention face's contribution is normally offset by the continuously increasing deflection force and the separation resistance continues to increase as illustrated in the three possible retention strength-deflection signatures shown in Fig. 3.32c. When calculating retention behavior, it is a good idea to calculate performance at partial release and just before final release. This will expose the situations where the angle effect dominates and the separation force drops once deflection begins.

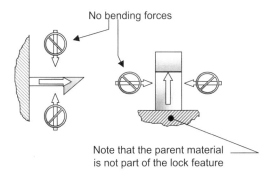

Figure 3.31 Locks should resist forces in the separation direction only

(a) Separation force initiates beam bending

(b) The retention face angle gets smaller as the beam bends

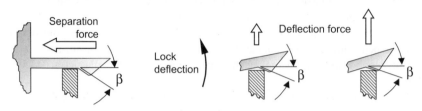

(c) The separation force-deflection signature is a function of decreasing angle and increasing deflection force and may be of increasing, constant or decreasing slope

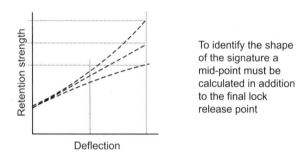

To identify the shape of the signature a mid-point must be calculated in addition to the final lock release point

Figure 3.32   Effect on retention strength as the beam deflects

Retention performance can sometimes be improved by adding a profile to the retention face, Fig. 3.33a. The profile compensates for the change in retention face angle and ensures that the instantaneous angle remains constant, Fig. 3.33b. This allows the lock to absorb more energy before releasing as shown in the force-deflection signature in Fig. 3.33c. When comparing force-deflection signatures, picture the separation energy absorbed as being proportional to the area under the curve. The signature can be modified for maximum effectiveness by adjusting the profile. The only limitations to the retention face profile are clearances for assembly and molding.

The retention face profile is a relatively subtle change and can be effective when forces are of very short duration, as might occur in a drop test of an electronic device. The principle of energy absorption can also be applied using some locators as spring-like features in the system. This is discussed in Chapter 4 in the section on compliance enhancements.

Another solution to preventing release is to use the loop style lock because it is inherently stronger both as a releasing and as a non-releasing lock. We have already discussed a non-releasing loop's retention behavior, but how will a releasing loop behave? Figure 3.34a shows a releasing loop and Fig 3.34b shows its separation force-deflection

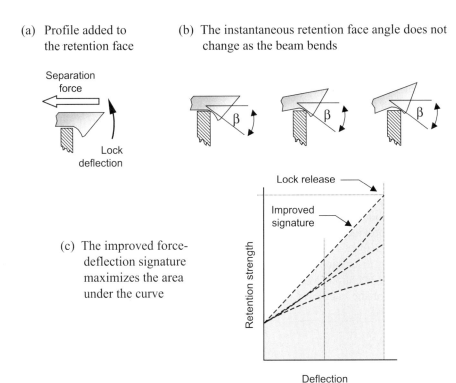

Figure 3.33 Benefits of a retention face profile

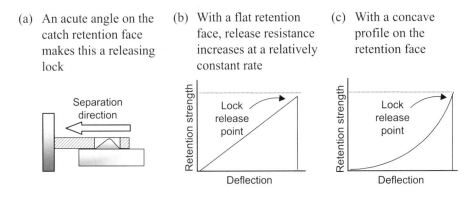

Figure 3.34 The retention force-deflection signature of the loop-style lock

signature. Even with a flat face on the catch, the retention force becomes greater as deflection occurs because the beam requires more force as it deflects. Adding a profile to the catch retention face will further improve energy absorption.

There are two more methods for making the cantilever lock stronger. One very effective and simple method is illustrated in Fig. 3.24g and Fig. 3.24i. The catch retention feature on the beam has simply been turned 90°. This change makes the hook's engage direction perpendicular to the long axis of the beam. This allows the beam to bend along the thin section for low assembly forces and low strain, yet resist separation across the thicker section of the beam's width. This is called "decoupling" and is discussed in detail in Chapter 5. Turning the catch 90° can also allow use of a cantilever lock where part clearance or mold design constraints prevent the use of a more conventional lock. The loop style lock can also be used in this manner.

The last method for improving cantilever lock performance is to add retention enhancements to provide additional support or strength to the lock. Retention enhancements are discussed in Chapter 4.

This concludes discussion of the cantilever style lock. It is the most common lock and deserves the most attention. Many of the fundamental principles for lock behavior were introduced here for the cantilever lock and will not be discussed again as the other lock styles are explained.

3.3.3 Planar Locks

Planar locks are so named because they are found on walls or surfaces (i.e. planes), Fig. 3.35. The walls are generally thin relative to their length and width so their behavior in deflection is plate-like. Therefore, the engagement and retention behavior of planar locks is described through the mechanics of plate deflection. A planar lock usually involves a catch on one part and an edge on the other part as the lock pair. One or both of the features may be on a deflecting wall on their respective parts.

These locks can be made relatively strong but, because at least one member of the lock pair must sit on a surface, the reaction force will always be off of the neutral axis. This creates the potential for distortion of the wall under high separation force and lock release.

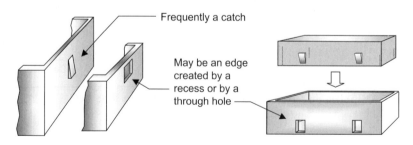

Figure 3.35 Planar locks

3.3 Lock Features   85

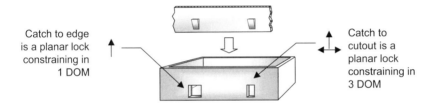

Figure 3.36 A planar lock can constrain in one or three degrees of motion

The weakness of the thin wall may require local support in the form of ribs or additional wall thickness in the area of the lock. An additional consideration with these locks is that both mating features can be on deflecting walls. This means both walls will deflect for engagement, reducing assembly forces and strains; but they can also deflect for separation, weakening the attachment.

The principles of the insertion and retention face profiles and their effects on assembly and separation force-deflection signatures are the same as for the cantilever hook. However, the more extreme deflections of the cantilever beam are not likely to be found in a wall.

Because a wall is likely strong in two axes, a planar lock can be made to constrain in three degrees of motion. In Fig. 3.36, the catch-cutout pair constrains in three DOM while the other constraint pair is a catch-edge. A second catch-cutout pair would not be appropriate here because it would create an over-constraint condition with the first.

3.3.4 Trap Locks

Traps engage through beam bending and retain through beam compression and/or bending so, like cantilever beam locks, trap behavior is based on beam mechanics. Traps differ from the beam locks, however, in both insertion and retention behavior, Fig. 3.37. A cantilever beam lock engages with the mating feature moving *toward* the fixed end of the beam and retains with the mating features moving away from the fixed end of the beam. Traps are just the opposite, engaging with the mating feature moving *away from* the lock base and retaining with the mating feature moving toward the lock base. These differences result in some significant performance differences between these two lock styles.

Traps can be extremely strong and are ideal as non-releasing locks for applications where parts are not intended for separation or where there is access from behind for manual release. With careful attention to the retention face contour, they can also be designed as releasing locks.

Traps seem to be quite common in solid-to-cavity, solid-to-opening and panel-to-opening applications where mating part removal is not expected. However, they are not limited to these applications, Fig. 3.38.

The beam shape in a trap is usually limited to variations of the straight cantilever beam. The retention mechanism is normally the end of the beam itself or a retention face formed by

(a) Engagement movement is toward the lock base

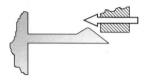

(b) Engagement movement is away from the lock base

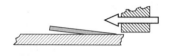

(c) Separation movement is away from the lock base

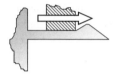

(d) Separation movement is toward the lock base

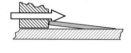

Figure 3.37 Cantilever beam vs. trap lock

(a) Releasing trap on four sides of a solid

(b) Non-releasing trap on a surface

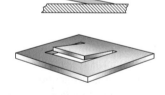

(c) Non-releasing traps on a tab

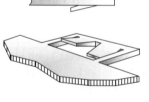

(d) A non-releasing panel-surface trap

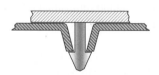

(e) Trap options for a solid-cavity application

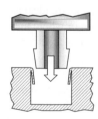

Figure 3.38 Trap lock examples

deformation of the beam. However, any form of the cantilever beam style lock can also become a trap when the direction of engagement is from the fixed end of the beam.

3.3.4.1 Trap Assembly Behavior

Because the trap lock's insertion face angle decreases and the point of contact enjoys an ever-increasing mechanical advantage as the trap is engaged, the assembly force signature shows a decreasing rate of increase, Fig. 3.39. This makes the trap an operator-friendly attachment because it tends to result in lower assembly forces and creates an over-center action for improved assembly feedback.

3.3.4.2 Trap Retention and Disassembly

Traps can be either releasing or non-releasing depending on the retention mechanism. Like the cantilever hook, a releasing trap resists separation through beam bending. Also like the hook, release behavior is a function of the angle and shape of the retention face and the coefficient of friction between the mating surfaces. There is one notable exception however. Unlike the hook, as separation occurs, the trap's retention face angle becomes steeper resulting in improved retention performance, Fig. 3.40.

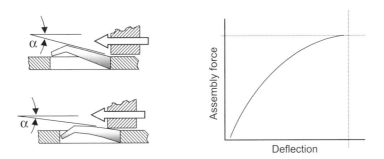

Figure 3.39 The trap lock assembly force-deflection signature

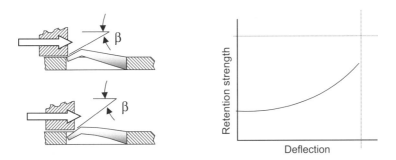

Figure 3.40 The trap lock retention strength-deflection signature

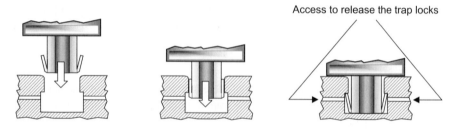

Figure 3.41 Non-releasing trap application

The non-releasing trap resists separation not through bending but through beam compression and can be very strong, Fig. 3.41. The failure mechanism of these traps is beam buckling and trying to force part separation will most likely damage the lock or the mating part. If applications using non-releasing traps are to be serviced, some provision must be made to allow access for trap deflection. A good example of the trap lock's strength can be found in the very common plastic tie-strap of the kind often used to bundle electric wires. The plain end of the strap is inserted into the locking end and pulled to engage the ribbed side of the strap with a ratcheting finger in the locking end. This is a trap mechanism and it is very strong.

Traps can also be found on some luggage buckles as an alternative locking mechanism to the cantilever hook. This kind of application (using a hook) is shown in Fig. 3.19d.

A non-releasing trap must ensure beam compression and protect against beam slippage and damage due to separation forces. (An exception is when the application is to be tamper evident and permanent damage to the lock is desirable or acceptable.) In the applications

(a) Separation force applied to an unrestrained trap lock

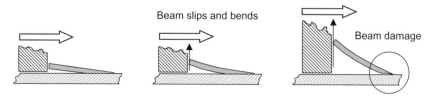

(b) Use of a tang to restrain a trap lock

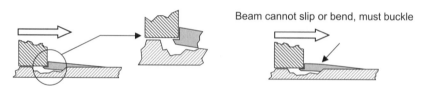

Figure 3.42 Non-releasing trap reaction to separation forces

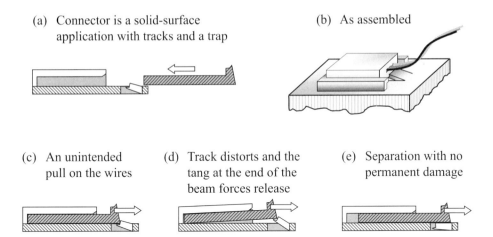

Figure 3.43 Non-releasing trap application

shown in Figs. 3.38d and 3.38e and in Fig. 3.41, the trap beams are prevented from slipping outward because they are contained by corners. The only possible failure is beam buckling which must occur at higher forces.

Fig. 3.42a shows the behavior of a non-releasing trap where beam movement is not restricted. As the separation force is applied, slippage may occur immediately if friction is insufficient. Or, once initial buckling occurs, the beam will slip on the mating surface. In either situation, resistance to separation forces is reduced and permanent damage is likely at much lower forces. Fig. 3.42b shows how a tongue or tang on the end of the lock beam can ensure against slippage.

Figure 3.43 shows a design (found on an electrical connector) where such an extension is used. We can assume the designer wanted a non-releasing trap to guarantee against accidental separation of the connector but also wanted to protect the trap from damage due to a hard pull on the wires leading into the connector. In this application, another interesting behavior was noted; it may or may not have been intentional. Because of the nature of this particular attachment, distortion can occur in the tracks when sufficient force is applied. When the distortion is sufficient, the tang pushes against the connector body and causes the beam to release. This behavior could prevent wire/pin separation in the connector.

A trap-like feature could also be added to a product strictly as tamper evidence and not as a lock. In this case it could be extremely small and need only to survive assembly deflection. Place it in a location where it is either visible before disassembly to ensure it is still in-place or deflectable with a tool if the disassembler knows where/how to deflect it. (Could "witness features" like this possibly become another group of enhancements?)

3.3.4.3 Traps and Lock Efficiency

Because lock feature design frequently involves a trade-off between assembly force and retention strength, a way of evaluating locks with respect to this trade-off can be very useful. The "lock efficiency" number provides a way of doing that. Lock efficiency is the ratio of a lock's retention strength to its assembly force [3]. Lock efficiency values can be developed for specific lock designs, but they are also useful for comparing the relative effectiveness of various lock styles. By its nature, the trap lock is inherently capable of developing the highest efficiency numbers of any lock style. As a general rule, if a trap lock can be used in place of a cantilever hook style lock, use it. The cantilever loop lock style is also more effective than the hook and has a higher lock efficiency rating.

$$E_L = \frac{F_R}{F_A} \qquad (3.3)$$

where E_L is lock efficiency; F_R is retention strength; F_A is maximum assembly force.

3.3.5 Torsional Lock

Torsional locks involve primarily torsional deflection for assembly although there is often some bending in the system as well. Retention depends on the stiffness of the torsion member and on the retention mechanism.

As shown in Fig. 3.44, the torsional member is not necessarily round. Torsional locks are relatively uncommon but are useful as an alternative to the cantilever style lock when clearances or access make hook location for disassembly difficult. For example, in an application where a hook must be flush with a panel and must be manually releasable (a non-releasing lock), the seesaw action of the lock allows release from the blind side of the retention mechanism.

Aside from the torsional deflection mechanism, the assembly and retention behaviors of these locks are similar to the cantilever beam style lock or to the trap, depending on the direction of mating feature engagement and release relative to the torsion member.

We are defining torsional locks as locks where the deflection mechanism is primarily torsion and can be analyzed as such. There are some locks where questions can arise as to their identification. In these locks, the installation deflection may be a combination of torsional shear, bending and plate deflection. Retention may involve torsional shear, plate deflection and either bending or compression. In these cases, evaluation of assembly and retention behavior will depend on which one of the deflection mechanisms dominates, and a thorough understanding of the interactions. An analysis which can evaluate these combined effects may be required.

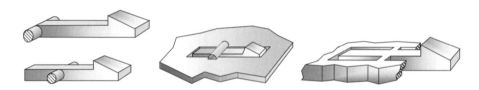

Figure 3.44 Torsional locks

3.3.6 Annular Lock

Annular locks involve interference between concentric ridges on cylinders and rely on radial elasticity for assembly and retention. Tensile and compressive hoop stresses occur in the lock features. An annular lock may be thought of as a catch wrapped around a cylinder and an edge wrapped around another mating cylinder, Fig. 3.45.

Note that, by this definition, a circular arrangement of hooks or traps is not an annular lock because it requires analysis of beam bending. Sometimes this arrangement is called "annular" in the literature. In that case, "annular" is a functional rather than a behavioral definition.

Annular locks can be extremely strong (permanent or non-releasing) or they can be releasing. A snap-on cap on a ball-point or felt-tip pen is a common releasing annular snap, the caps on 35 mm film canisters are another. They can also permit (free) rotation in a moveable application. Because annular locks, by definition, involve a locator pair (pin-hole, Fig. 3.44a) or the entire mating part-base part system (solid-cavity, Fig. 3.44b) they will constrain in more than one degree of motion. Normally, annular locks constrain in 5 DOM. This is another difference from the circular hook or trap arrangement.

3.3.7 Lock Pairs and Lock Function

In Chapter 2, we introduced a descriptive element of a snap-fit called *function*. Function describes what the locking feature(s) in a snap-fit attachment must do. Locking feature function is explained in terms of *action*, *purpose*, *retention* and *release*. These generic descriptions of locking requirements help us to categorize snap-fit applications for benchmarking and organizing snap-fit application libraries.

(a) Annular lock as a pin-hole locator pair

(b) Annular lock as part of a solid-cavity application

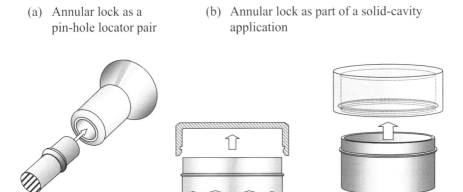

Figure 3.45 Annular locks

Now that lock features have been defined, the reader should be able to see that, depending on the design requirements and limitations, certain lock features and lock pairs will be preferable to others for a given lock function. Likewise, a lock pair can be over-designed or over-engineered causing unnecessary engineering time and manufacturing cost. It is important to balance the application requirements with the cost/capability of the various lock options.

3.4 Summary

The purpose of this chapter has been to present a descriptive explanation of the various constraint features. From this information, a designer should understand the fundamental differences in constraint feature behavior and be able to select the appropriate constraint feature styles when developing an application concept.

This chapter described the two major kinds of constraint features, locks and locators, which are used in the snap-fit interface to create a constraint system. Constraint features remove degrees of motion from the attachment and are the "necessary and sufficient" conditions for a snap-fit attachment.

3.4.1 Important Points in Chapter 3

- The fundamental problem in snap-fit design is that locks must be weak in order to deflect for assembly yet strong enough to prevent part separation.
- Snap-fit reliability depends on establishing and maintaining a line-to-line fit between the mating and base parts. Do not expect to get any significant or long-term clamp load in a snap-fit.
- The rules for mechanical advantage and dimensional robustness that were introduced and explained with the locator features are general rules for all constraint features and they also apply to lock pairs.
- Some locator pairs can constrain in as many as 5 degrees of motion, others in as few as one. Most lock pairs, planar and annular being the exceptions, can constrain in only one degree of motion (the separation direction). Designing for a lock to constrain in additional degrees of motion will leave the attachment under-constrained.
- The cantilever beam, planar and trap are the most common lock styles. Torsional and annular are often special usage locks.
- A profile added to both the insertion and retention faces in a lock pair can significantly improve assembly and retention performance.
- Caution, even non-releasing hooks will release under sufficiently high forces.
- Locators can be used as low-deflection locks, particularly when an assembly motion that involves sliding (slide, twist and pivot motions) is present.
- Lock efficiency, the ratio of retention strength to assembly force, is a good indication of inherent lock effectiveness.

3.4.2 Design Rules Introduced in Chapter 3

- Because of the tendency of plastic to creep, avoid long-term or sustained forces across the snap-fit interface unless these forces are low and long-term performance is indicated by analysis and verified by end-use testing.
- Use locators to carry all significant transient forces across the interface and arrange locks so they do not carry transient forces in the separation direction unless they are permanent locks or have special retaining capability.
- Locators should be the first constraint features added when developing the snap-fit interface and the first locators considered should be the one(s) that make first contact during assembly. This locator pair(s) should also provide the guidance function.
- Because locators are strong relative to locks, the more degrees of motion that can be removed with locator pairs in a snap-fit, the stronger the attachment.
- The assembly motion selected for an application will determine the potential strength of a snap-fit attachment. This is because locator pair selection for a given application is a function of the assembly motion.
- The potential for degrees of motion removed by locators is highest with the tip, slide, twist and pivot motions so they are generally preferred over the push motion. Of the four preferred motions, tip is usually the most practical.
- When a beam lock is being considered, design to use loops or traps whenever possible. In general, the hook style locks will have the lowest lock efficiency. Cantilever beam loop style locks have much better efficiency than the hook, and the traps have the highest.
- A loop or trap lock used with a tip assembly motion is a highly effective snap-fit attachment concept and should always be considered as a design alternative.
- Where two natural locators make up a position-critical pair and fine-tuning may be necessary, consider adding a discrete locator as one member of the pair.
- Maximize part stability and minimize sensitivity to dimensional variation by placing constraint pairs constraining the same rotation or translational movement as far apart from each other as possible.
- Maximize mechanical advantage against rotational forces by placing constraint pairs acting as a couple so that their (parallel) lines-of-action are as far apart as possible.
- Specify a radius on all interior and exterior corners of constraint features. This applies to the feature intersection with the parent material and to all corners within the feature itself.
- Non-releasing trap locks must be protected from over-deflection and damage.
- Design compensation for knitline weakness into loop style locks.

References

1. Loops were described as a unique lock feature in *Integral Fastener Design*, Dave Reiff, Motorola Inc., Fort Lauderdale, FL.

2. Plastic Part Design for Economical Injection Molding, 1998, Glenn L. Beall, Libertyville, IL. Test data reproduced from LNP Cloud, McDowell & Gerakaris, Plastic Technology, Aug. 1976.
3. Luscher, Dr. A.F., *Design and Analysis of Snap-fit Features*, from the Integral Attachment Program at the Ohio State University, 1999.

4 Enhancements

Enhancements were introduced in Chapter 2 as the second of two groups of physical elements used in snap-fit attachments. They were also referred to several times during the discussion of constraint features in Chapter 3. In this chapter all the enhancements are presented and described in detail.

4.1 Introduction

Enhancements may be distinct physical features of an interface or they can be attributes of other (physical) interface features. They improve the snap-fit's robustness to variables and unknown conditions in manufacturing, assembly and usage. In other words, they make a snap-fit more "user-friendly". Most enhancements do not directly affect reliability and strength but, by improving the snap-fit's robustness to many conditions, they can have very important indirect effects on reliability. They are a big part of the "attention to detail" aspects of good snap-fit design.

Enhancements are often tricks-of-the-trade that experienced snap-fit designers have learned to use. Meanwhile, the inexperienced designer must learn their value through trial-and-error. Read this chapter thoroughly. Enhancements will do more for your application than you can imagine.

A snap-fit application does not require enhancements. Only constraint features are absolutely necessary in a snap-fit attachment. But, as we will see, enhancements are required if a snap-fit is to be "world-class". If you have examined some snap-fits and found features you could not identify, or maybe wondered, "Why did they do that?" you may have been looking at an enhancement. If you have assembled and disassembled similar snap-fit applications from different sources and marveled at how such similar applications could behave and feel so different, credit the difference to enhancements. As consumers and users of plastic products, we regularly use enhancements. If you have been frustrated by a snap-fit, chances are it was because of improper use or lack of enhancements in the product.

Certain enhancements should be considered as requirements in every application. Others are required depending on the nature of the application. Still others can be thought of as "nice-to-have" but not essential.

Benchmarking is an important part of the creative process for snap-fits and the subject of enhancements and benchmarking deserves special comment. As you conduct technical benchmarking studies of products, many of the best ideas and creative hints will not be dramatic or highly interesting product features. They will be subtle and rather mundane details in the parts; much like those described in this section. By studying the enhancements on parts, you can find important clues to the problems the product designers had to overcome and how they did it. You can then predict and avoid problems of your own. Benchmarking is discussed in more detail in Chapter 7 as a part of the snap-fit development process.

Enhancements are grouped into four categories according to their effects on the attachment: *assembly, activation, performance,* and *manufacturing*.

4.2 Enhancements for Assembly

Assembly enhancements are features and attributes that support product assembly. They help to ensure that the assembly process will consistently and efficiently produce a good attachment. Two kinds of assembly enhancements are identified, *guidance* and *feedback*. Both are required in all snap-fit applications.

Imagine a worst case scenario for assembling snap-fit parts; it might require an operator to perform six steps. The first five are addressed by guidance enhancements, the last by feedback.

- Initial alignment—Gross movements to orient parts for engagement.
- First adjustment—Small motor movements to engage the first locators.
- Second adjustment—Small motor movements to engage additional locators and overcome minor feature interference as parts are moved to final locking position.
- Third adjustment—Small motor movements to align locks.
- Locking—A force is applied to engage the locks and complete the attachment.
- Verification—The operator is satisfied that a good attachment has been made.

Each of these steps takes time. Also, whenever extra or unnecessary movements occur, they have the potential of contributing to cumulative trauma injury. Assembly enhancements can simplify the assembly process to:

- Initial alignment—Gross movements to orient parts for engagement.
- Engagement—Small motor movements to engage the first locators.
- Locking—A force is applied to engage the locks and complete the attachment.
- Verification—The operator is satisfied that a good attachment has been made.

4.2.1 Guidance Enhancements

Some of the guidance examples will seem trivial and readers may say to themselves, "I would never do anything like that!" The truth is these kinds of design oversights can be found on numerous products. They are rarely so dramatic or obvious as to attract a lot of attention; they just accumulate in the design details, adding cost to the product by reducing productivity. These are also the kinds of design flaws that disciplines such as design for assembly and design for manufacturing try to eliminate because the cost penalty for a difficult-to-assemble design can be significant.

(a) Pins and posts as guides   (b) Guides as extensions on other features

Figure 4.1 Guide features

Initial mating part to base part alignment followed by a simple assembly motion (push, slide, tip, twist or pivot) should be all that is necessary to fasten parts. An operator should not need to struggle or make small adjustments to align the mating part to the base part to initiate assembly. Once assembly is started, the mating part should locate itself to the base part and require only a final push by the operator to complete the attachment. This is the role of guidance enhancements.

Guidance is broken down into features called *guides* and *pilots* and an attribute, *clearance*. Guides and clearance enable ease of assembly. Pilots ensure that parts susceptible to incorrect assembly orientation are properly installed.

4.2.1.1 Guides

Guides help the assembly operator by simplifying the gross movements required to carry out initial engagement of the parts. Guides stabilize the mating part to the base part so the operator can easily bring the parts together without feature damage and without wasted time or extra movements.

Some common guide features are shown in Fig. 4.1. Note that some of the guides look exactly like locators. The guide function may be carried out by a distinct guide feature dedicated to that purpose, but it is usually more efficient to carry out the guide function by using a feature that already exists in the interface. Most of the time, the guide function can and should be incorporated into selected locators. When locators are to be used as guides, add the guide function to the first and, if necessary, the second locator pair(s) to be engaged. Recall this was also discussed in the locator feature section of Chapter 2.

In some situations, where precise alignment of locking features is required for ease of assembly, guides should also be built into the lock pairs. This may be necessary if a lock feature or the wall on which it is mounted is subject to some warping and its final position is somewhat variable.

Some general rules for guide usage:

- Lock features should never be the first features to make contact with the other part, Fig. 4.2a.
- For ease of assembly, guides must engage before the operator's fingers contact the base part, Fig. 4.2b.

(a) Without guide features, the operator must make fine adjustments to align the part and the locks are susceptible to damage

(b) Guide features (pins) align the locks with the edges; no fine adjustments are needed and the locks are protected

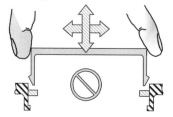
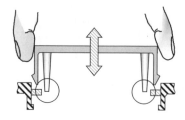

(c) When multiple guides or locators engage holes and slots, one must engage first to stabilize the part

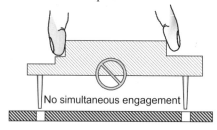
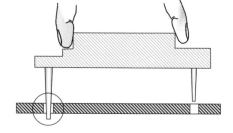

Figure 4.2 Guide feature usage

- For ease of assembly, avoid simultaneous engagement of multiple guides. One or two guides should engage first to stabilize the mating part to the base part, Fig. 4.2c. This is particularly critical when the guides are protruding features engaging into holes or slots. It is less critical if the guides are engaging against edges or surfaces. This is also a good rule to follow with respect to locators.
- A "tip" assembly motion can eliminate or reduce simultaneous engagement because it forces initial engagement at one end of the part followed by rotation to sequentially engage the remaining features.
- Build the guide function into existing constraint features whenever possible.

4.2.1.2 Clearance

Once the mating part is stabilized to the base part by guide features, clearance attributes ensure all features in the interface (including guides) can be brought together without interfering or hanging-up on each other, Fig. 4.3. As with guides, wasted motions are eliminated, this time because minor part position adjustments are not needed.

Clearance is not difficult. It is simply thinking about all possibilities for part-to-part interference and eliminating them. In general, clearance is achieved by designing generous

(a) In a solid to cavity or opening application: specify a radius or bevels at all initial contact points and design for clearance between the parts for initial engagement

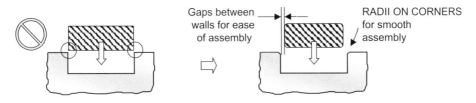

(b) In a track locator, replace sharp corners with radii or bevels at all initial contact points

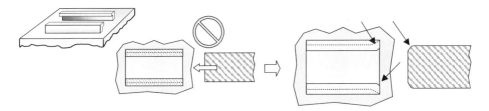

(c) Use tapered features and replace all sharp corners with a radius

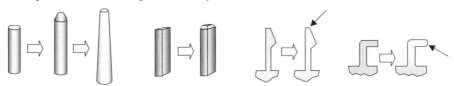

Figure 4.3 Clearance is an attribute of both features and parts

radii on all edges and by tapering the locators and guide features. This is a very simple concept but it is often overlooked in practice. Some clearance rules are:

- Always specify a taper or a radius on all corners and edges of the parts proper as well as on all the features. This is also an important requirement for proper mold design.
- Always provide generous clearance for initial engagement, again on the parts proper as well as the features.

4.2.1.3 Pilots

Pilots are used to ensure proper orientation of a mating part that may otherwise be assembled incorrectly. This is the case with symmetric parts that can be assembled more than one way. Pilots may be distinct features arranged to allow one-way assembly, Fig. 4.4. Or, as in Fig. 4.5, guides or locators can be made to perform the pilot function through asymmetric arrangement. This avoids the cost of adding a special pilot feature.

(a) Switch design A

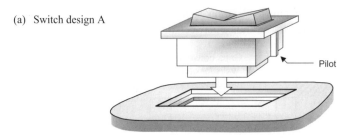

(b) Switch design B

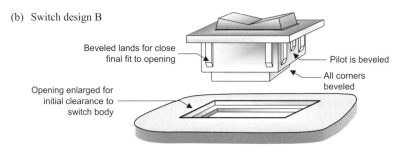

Figure 4.4 Switch application without and with guidance features

(a) Original design has ten interface features, the part is a bezel ~ 50 mm x 50 mm

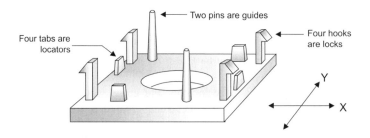

(b) Redesign has six interface features

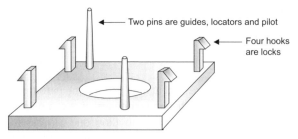

Figure 4.5 Guidance, constraint features and efficient design

4.2.2 Product Example 1

Consider the rocker switch application shown in Fig. 4.4a. This (solid to opening) example is based on a real application. Design A has no guidance enhancements at all and the time to get and install it was measured at 7 seconds.

Several factors contribute to the assembly difficulties with this application. The walls of the switch body are acting as locators to the edges of the opening. To prevent relative movement after it is assembled a line-to-line fit is required. This provides no clearance for initial engagement and the sharp corners on both the solid and the opening make engagement even more difficult. The mating part is also unstable because the operator must hold it by the moveable rocker switch while trying to find the line-to-line fit required for initial engagement. The nature of the switch design and the styling around the opening force this last condition but that is even more reason to make the design easy to assemble.

An improved switch body is design B shown in Fig. 4.4b. It is from the same kind of application but from a different supplier. This design makes good use of guidance principles and the time to get the part and install it into the opening is 3 seconds. Relief is provided for easy initial engagement by over-sizing the opening relative to the switch body. Once initial engagement of the parts occurs, the required line-to-line fit is obtained through use of lands as locators on each wall. Beveled faces on the lands and around the opening and leading corners of the walls provide additional clearance so no additional small motor movements are required. A pilot feature, also with a bevel, ensures correct switch orientation in the opening.

The time difference between these two designs is "only 4 seconds". Nevertheless, over time, the cost in assembly time can become significant. In Table 4.1, the estimated cost of 4 seconds of wasted time is shown for several labor rates and part volumes.

Table 4.1 Cost of Four Seconds of Assembly Time per Unit

Units per year	Labor rate $/hr				
	8	10	15	20	25
20,000	176	200	340	440	550
50,000	440	550	823	1100	1373
100,000	880	1100	1647	2200	2747
200,000	1760	2200	3294	4400	5494

Other costs, like burden, could be added into these numbers, but there are also additional problems and costs that could be associated with design A. These other costs may be difficult or impossible to measure but have the potential to be much higher than the assembly time cost alone. Operator frustration as a result of struggling to assemble the parts might result in quality problems and, regardless of the quality aspects, operator frustration in itself is undesirable. The extra finger and wrist movements required during installation might result in the added cost of workers' compensation for cumulative trauma injuries. If the product is intended for automatic or robotic assembly, higher cost equipment might be needed to get the precise control required to assemble the parts. In this case, the designer would likely look for ways to reduce the assembly precision required. If one would try to design this product to be easy for a robot to assemble, why not design it to be easy for a human being?

The point is that for just a little more effort in design and little or no increase in piece cost, a product that is much easier to assemble can be designed. Thinking about guidance in terms of robotic assembly is not a bad idea:

"If you want to learn how to design products for people to assemble, hang around with robots." [1]

4.2.3 Product Example 2

Another example of guidance is shown in Fig. 4.5a. This relatively small (50 × 50 mm) and low mass bezel probably does not need ten distinct features to do its job. We cannot know the exact reasons for this design; perhaps ten features are indeed necessary but, for the sake of this discussion, we will assume they are not needed. We will suggest some changes to make the attachment a little more efficient. Without seeing the base part, we also cannot know if this attachment is over-constrained in the x–y plane, but there is a good chance that it is.

Possible changes to the part are shown in Fig. 4.5b where the suggested redesign uses four fewer features. The pins are used as both locators and guides and are different lengths for sequential rather than simultaneous engagement. A taper on the pins allows easy initial engagement and, once seated, the pins will have a line-to-line fit with the mating locators. A hole and a slot will be needed in the base part to accommodate the locators and we can now be certain the application is not over-constrained. The original design had good clearance attributes, radii, bevels and tapers on all the features and these are also used in the redesign. The original design also had a pilot function through asymmetric arrangement of the pins; is also carried over in the new design.

4.2.4 Product Example 3

This example is based on a real product problem that required some investigation to determine the root cause of the problem before it could be fixed. It is an excellent example of how proper use of enhancements can improve several aspects of snap-fit performance. As

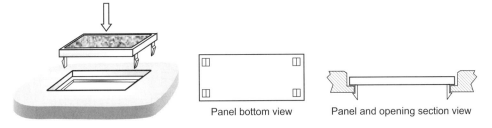

Figure 4.6 Example panel-opening application, original design

often happens, the simplicity of the parts most likely caused the original designer to consider the application as "easy". The result was a poor snap-fit and the expense of fixing it.

The application is a panel-opening application in which a small plastic panel (the mating part) attaches to an opening in a large panel (the base part), Fig. 4.6a. The mating part as originally designed used four hooks as locking features. Panel-opening applications are a common snap-fit design situation and cantilever hooks are often used as the locking feature in this kind of application.

The problem with this application was that, in some products, the small panel (about 30 by 80 mm) would fall out of the opening in a relatively short time after assembly. Customers were, of course, disappointed that such a simple attachment could fail and the part had to be replaced under warranty.

At first glance, the cause of the problem appeared to be lock feature failure because returned parts always had one or more broken hooks. A traditional (and logical) conclusion would have been that the hooks were weak. The solution would have been to design stronger hooks. However, an attachment level diagnostic approach is to look at the application interface as a system before reaching any conclusions. (Diagnosing snap-fit problems is covered in Chapter 8.) By thoroughly examining the application for systemic problems before simply fixing the hooks, we find that several enhancement-related aspects of snap-fit design must be fixed before addressing the hooks themselves.

To properly evaluate any snap-fit problem, one must get parts and observe the assembly operation itself. (Ideally, you will have the opportunity to actually assemble the application in the production environment.) Without parts to "play with", you will not be able to understand the problem properly. This application is no exception.

In trying to assemble the original design parts, we find that, with a normal grasp of the part, the fingers contact the base part before any constraint features on the mating part can engage, Fig. 4.7a. The operator cannot properly hold the mating part to align the hooks before trying to push it into its locked position. Maybe the operator is sometimes pushing the part into the opening when the hooks are not lined up with the edges of the opening and this is damaging the hooks so they are ineffective in holding the part in place. Guides can be added to extend far enough into the opening to provide initial alignment and ensure the hooks are aligned with the edges before the operator's fingertips interfere with the base part. We also realize that the operator's vision of the opening is partially blocked by their hand during assembly. This is another good reason to have effective guide features in this application.

(a) Finger interference occurs before the hooks can engage and the operator's hand interferes with their view of the area, making it a blind assembly

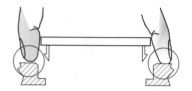

(b) Guide features eliminate alignment problems and hook damage during assembly

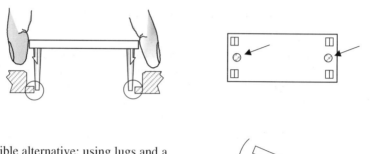

(c) Possible alternative: using lugs and a tip motion eliminates the need for separate guide features

Figure 4.7 Example application, improving ease of assembly

When we add the guides, Fig. 4.7, and try out the new parts, we find the guides do orient and stabilize the part. However, hook failures are still occurring, but at a lower rate. Something is still wrong, but this does not mean the guides were a bad idea; guides are never a bad idea. It just means there is more than one problem with this design.

Figure 4.7c shows another possible design for this application. As a rule, when access and part shapes permit, as with this example, the tip assembly motion with a lug(s) at one end is always preferable to the push motion.

We will come back to this application after learning about operator feedback.

4.2.5 Operator Feedback

Feedback is the second assembly process enhancement. When operators assemble snap-fits, their hands are the assembly tools. Unlike operators using power tools that shut off at a specified torque or robots with sensors, the snap-fit assembler has no calibrated tool or electronics providing indirect feedback when a good assembly has been made. The snap-fit assembler has something better; their sensitive fingers, eyes and hearing all connected to a powerful processor, the human brain. The operator relies on direct feedback from the assembly process to indicate success of the assembly. Designing the snap-fit to ensure consistent and positive feedback to the operator helps ensure that properly assembled attachments occur every time.

The goal during snap-fit design is to improve and amplify direct feedback to the operator while eliminating or minimizing other factors that can interfere with the direct feedback. We can think of these interfering factors as "noise" in the system. Direct assembly feedback has three forms: *tactile*, *audible* and *visual*.

Tactile feedback results from the sudden release of energy, usually the lock(s) snapping into place. It is enhanced by the shape of the assembly force-deflection signature and by the solid feeling created when the locator pairs come together. Tactile feedback is generally preferred over the other forms of feedback because it is not subject to audio or visual interference.

Audible feedback is also the result of a sudden release of energy. Ambient noises and possible operator hearing limitations may reduce its effectiveness.

Visual feedback involves alignment of visible mating and base part features. Plant lighting, line-of-sight interference and operator limitations may reduce its effectiveness. It may also require a subjective judgement on the part of an operator or inspector.

Ideally, more than one source of feedback should be available to the operator. The sudden release of energy that gives a tactile signal may also cause an audible signal. Position indicators may provide a visual indication to supplement an audible or tactile signal.

Tactile feedback can be understood if we think in terms of the assembly force-deflection "signature" that was introduced during the lock discussion in Chapter 3. The signature represents what the operator feels as the mating part is installed. Some common snap-fit assembly signatures are shown in Fig. 4.8 along with some of the lock insertion face contours that produce them.

The concave curve in Fig. 4.8a is typical of many attachments. It has a geometrically increasing force as the insertion face to mating surface contact angle increases with (cantilever) hook deflection. The parts then make solid locator contact as the lock(s) engage. In many cases, this is acceptable and provides adequate feedback but in applications with high feedback interference, it may not be sufficient.

Improved feedback and assembly feel occur when the insertion face profile results in either a flat or a convex signature Fig. 4.8b. The maximum assembly force is generally lower for the same deflection, which means that lock deflection may be increased for a stronger signal. (Remember that strain limits in the lock material must also be considered before increasing beam deflection.) A flat signature is produced when the instantaneous insertion face angle remains constant with respect to assembly deflection. A convex signature is produced when the instantaneous insertion face angle decreases with respect to the assembly deflection and is inherent in the trap lock feature. A discussion of insertion face contour can be found in Chapter 3 and some analysis principles are presented in Chapter 6.

The signatures shown in Fig. 4.8c represent applications where soft materials in the interface, structurally weak components or compliant locators may require the operator to hunt for engagement because the locator contact and lock engagement points are not well defined.

Similar feedback issues exist in moveable applications where the customer operates the snap-fit. These are discussed under the enhancement topic called "user-feel".

(a) Typical assembly signature

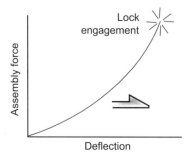

(b) Possible assembly signatures

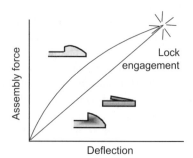

(c) Soft or compliant parts or weak constraint features

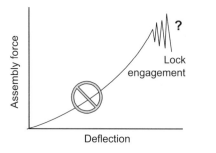
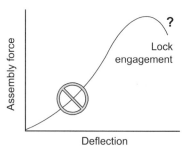

Figure 4.8 Tactile feedback assembly force-deflection signatures

Good feedback is generally obtained by adjusting the attributes of existing part features. Note that most of the design characteristics that support good operator feedback are related to tactile feedback.

Ergonomic factors also affect operator feedback. Assembly forces must be within an acceptable range. A comfortable operator position, normal motions and parts that assemble easily will help create a work environment in which the operator can be sensitive to tactile feedback. Some general ergonomics rules are:

- Avoid extreme rotational or reaching motions.
- Avoid high forces on fingers, thumbs or hands to install a part. High cumulative assembly forces (as multiple locks are engaged) can interfere with feedback as the operator struggles to overcome them. Lock designs to reduce assembly forces are discussed in the chapter on lock features.
- Avoid awkward reaches or twisting motions.
- Avoid reaches over the head.
- Design for top down, forward and natural motions carried out from a comfortable body position.

Other factors that support improved operator feedback include:

- Provide solid pressure points. If a mating part is compliant, stiffen the points at which the operator must apply pressure to locate and lock the part in place. They must be structurally sound to transmit force to locators and locks with little or no deflection. Weak parts or soft materials may require local strengthening.
- Positive and solid contact between strong locator features will send a clear, unmistakable signal that parts are positioned properly against each other.
- A rapid lock return can give a good audible and tactile signal that the lock is engaged. A lock with high deflection is generally more effective than one with low deflection. High deflection does not necessarily mean high assembly force, however. The audible feedback signal is generated by lock feature speed as it snaps into place, not by lock force.
- A strong "over-center" action as a lock engages will give a feeling that the part is being pulled into position.
- Consistency in part assembly performance allows the operator to acquire a feeling for a good attachment. Once this feeling exists, anything out of the ordinary will signal the operator to check for problems. Consistency of performance is a function of the design's robustness to manufacturing and material variables.
- Provide highly visible features that are clearly aligned when the assembly is successful.
- Design for go/no-go latching. This means that a part that is not properly locked in place will easily fall out of position to create an obvious assembly failure that can be fixed immediately.

Poor operator feedback is caused either by poor execution of the characteristics that provide good feedback or by failure to eliminate the background "noise" that interferes with feedback. Causes of poor feedback include:

- Compliant components and soft materials that flex and bend so part position is in doubt.
- Soft materials and low deflection locks that do not release enough feedback energy when the lock engages.
- High forces or assembly forces of long duration such that the operator's fingers lose their sensitivity to tactile feedback.
- False assemblies that look good immediately after assembly to fool the operator and inspectors but fail later.
- Inconsistent assembly behavior. Parts that lack consistency during assembly make it difficult for an operator to develop a feeling for a good attachment.
- Awkward positions and motions. Anything about the assembly operation that is poor from an ergonomic standpoint will interfere with tactile feedback and, in any case, will make it more difficult for the operator to do a good job.
- Difficult assembly. Anything that creates a difficult assembly operation will interfere with the operator's ability to recognize a poor attachment if it occurs. Using guide enhancements to make the assembly process as easy as possible will reduce system "noise" and improve the quality of the feedback.

108   Enhancements   [Refs. on p. 134]

4.2.6 Product Example #3 Revisited

Let us now go back to the application problem involving the small panel, Figs. 4.6 and 4.7. The hooks in the original design were sometimes damaged or broken during part insertion and could not engage properly. After guidance features were added, the lock failures continued, but at a lower rate. We discover that a soft covering on the base part can also affect the edge thickness where the hooks engage. Sometimes, the soft covering is not well trimmed and can wrap around the edges of the opening. Sometimes, even when properly assembled, one or more hooks may not fully engage. The soft covering and short (low deflection) hooks are preventing any positive feeling of part seating and engagement and the operator receives no tactile feedback of proper locking. The assembly signature looks something like that those in Fig. 4.8c. But, even a part with broken or damaged hooks could remain in place, appearing to be properly assembled, for a while. The poorly designed hooks are partially responsible for the difficult assembly and other factors make it hard for the operator to identify poor assemblies. Thickness variation in the material around the edge of the opening makes hook engagement unreliable. A new lock design that will provide better feedback to the operator is needed. A lock that is less sensitive (more robust) to edge thickness variation is also desirable but that is a lock feature design issue, not an enhancement. The redesigned attachment [2] is shown in Fig. 4.9 where:

- The guide features are now tabs that carry the (trap) locking features.
- The longer lock beams allow greater deflection for higher feedback energy.
- The new lock style (trap) is more tolerant of the edge thickness variation.

To summarize for this example, the enhancement-related problems were:

- No guidance, which resulted in difficult assembly and damaged hooks.
- No feedback, resulting in poor and damaged assemblies going out to customers.

These problems affected both ease of assembly and attachment reliability although the locks themselves were strong enough to hold the mating part in place. Lock related problems were:

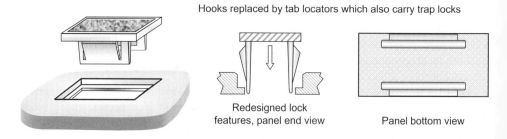

Figure 4.9 Lock feature changes for acceptable engagement and operator feedback

- The extremely short hooks caused high assembly force, were inherently susceptible to high strain even when properly aligned during assembly, and were highly susceptible to damage if not properly aligned during assembly.
- High assembly force but low deflection, generating no tactile feedback energy.
- No tolerance to thickness variation at the edge of the opening.

Obviously, just making the locks stronger would not have solved all the problems with this attachment. In fact, just making the hooks stronger may have made the problem worse because assembly effort would have increased.

In this particular application, there was enough depth in the opening to allow use of the deep guide/locator features and the trap style lock. Sometimes, due to clearances, we do not have the luxury of unlimited space. There are other ways the lock features in this attachment could have been designed to solve the problem, one possibility is the side-action style lock as shown in Fig. 3.24g. This kind of side-action lock is ideal for use in limited spaces.

4.2.7 Assembly Enhancements Summary

Many of these assembly enhancements should be familiar to those acquainted with design for assembly principles. They are extremely important because they can reduce assembly time and because, by making assembly easier, they help to ensure that a good attachment is made every time.

Guidance is ensuring smooth engagement and latching of mating parts. This topic is further broken down into guides, clearance and pilots.

Operator feedback involves attributes and features to ensure clear and consistent feedback that the attachment has been properly made.

4.3 Enhancements for Activating and Using Snap-Fits

Activation enhancements are mechanical and informational features that support attachment disassembly or usage. These enhancements make it easier to use a snap-fit application. Most of the time, activating a snap-fit means releasing it, either to separate parts or to operate a movable snap-fit. In the case of a movable snap-fit, activation can also mean re-locking the attachment after use. Enhancements for activation are *visuals*, *assist*, and *user-feel*.

Visuals provide information about attachment operation or disassembly. Assists provide a means for manual deflection of non-releasing locks. User feel refers to attributes and features that ensure a high quality feel in a moveable snap-fit.

4.3.1 Visuals

Sometimes the operation of a snap-fit is obvious. When operation is not obvious, visuals provide a message or indication to the user of exactly how to use the snap-fit. Visuals make the snap-fit easy to use and help prevent damage due to misuse. Examples of common visual enhancements are the arrows on battery covers of most television and VCR remote controls. Many children's toys have visuals indicating how to open, move or remove parts. Visuals may also be instructional text located close to the attachment's activation point.

Recall that part separation is accomplished by reversing one of five simple assembly motions, (push, slide, tip, twist or pivot). Thus, a simple visual indication of the mating part's separation direction and motion may be sufficient when the application uses a releasing lock. When a lock feature is non-releasing, both an indication of the manual deflection to release the lock and an indication of mating part separation motion may be necessary.

Examples of some common visuals include arrows on battery covers (on toys and remote controls) indicating how to remove the cover, instructional text on non-appearance surfaces that describes the disassembly operation and thumb depressions accompanied by a directional indicator.

Visuals should be large so they are easy to find and understand when they are in an area of the part where appearance is less important. A visual on an important appearance surface, however, cannot be obtrusive or unattractive yet customers and service personnel must be able to find it and interpret its meaning. As snap-fits become increasingly common in products, users (both consumers and service personnel) must learn to expect and look for visuals.

In place of, or sometimes in addition to visuals on the parts themselves, text instructions can be given on nearby labels or in product owner's and service manuals. A visual pointer to these instructions may be appropriate.

The primary purpose of visuals is to avoid part and feature breakage during the useful life of the product. However, material recycling and reuse once the product's useful life is over are also becoming an important product concern. The trend toward "green design" is moving strongly across the world and should not be ignored. Parts not intended for disassembly during the useful life of the product must still be efficiently disassembled for material recycling or salvaged for reuse. When part disassembly is not obvious, visuals can indicate a breaking point or a critical point for efficient part separation. The common recycle symbols that use a number indicating the family of material for separation and reprocessing are visuals that support recycling.

When designing snap-fit visuals, keep the customer in mind. Locking and releasing methods should be as obvious as possible and the supporting visuals intuitive and readily visible. Remember that the typical customer will be totally unfamiliar with the parts and the attachment method and even experienced service technicians will need to become familiar with new designs.

While standards exist for many symbols, no set of standard international snap-fit symbols has been identified. In addition to the more recognizable visuals like text and arrows, certain cryptic visuals are needed for use in limited space areas or in appearance areas where large obtrusive marks would be unacceptable. Industry leaders in plastic products should take steps to establish an international set of standard symbols.

As a starting point for such an initiative, some possible symbols are shown in Fig. 4.10. These shapes are proposed to describe snap-fit activation (release or operation) when space or appearance considerations prevent more detailed information [3]. Standards for symbol

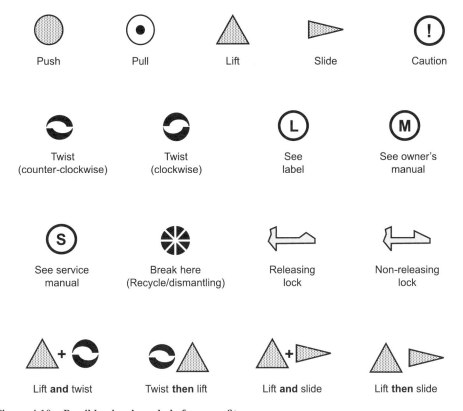

Figure 4.10 Possible visual symbols for snap-fits

geometry exist and should be applied to determine actual symbol dimensions. The SAE Recommended Practice J1344 describes a system for marking plastic parts with material identification symbols. The SAE system is based on the standard symbols for plastics [ISO 1043] published by the International Organization for Standardization. The SAE system indicates text letters 3 mm in height. It is possible that symbols could be smaller than 3 mm and still be identifiable.

4.3.2 Assists

Assists are the second enhancement feature that helps with activating the snap-fit. Assists make it easier to release a locking feature or operate a movable snap-fit. When lock operation is hidden or not obvious, an assist should be accompanied by a visual to indicate how the assist is to be used. Again, showing the user how to release the lock may prevent product damage.

Some examples of assists are shown in Fig. 4.11. Finger activation of the lock feature using the assist is preferred, but sometimes tool activation is necessary. Any form of manual lock deflection requires caution because over-stressing the lock feature can be very easy to do. This is particularly true if a tool is used and/or if the lock feature is in an area that is difficult to see or reach. Guard enhancements can help prevent lock damage due to over deflection and are sometimes used with assists. Guards are described in an upcoming section.

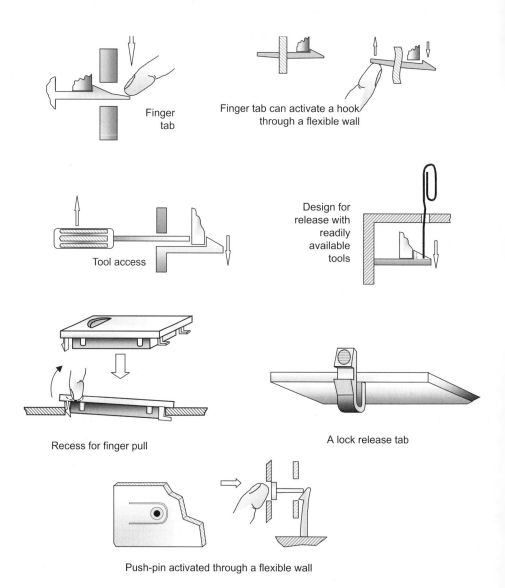

Figure 4.11 Examples of assists

Assists may also be used for part assembly, for activating a movable snap-fit or for assisting part movement to unlatch a releasing lock. An access opening can even be skinned over with an indication (a visual) on a visible surface to drill or punch through at that point to reach the lock. A surface or exterior operated assist feature can be used to activate a lock buried in the interior of a part. Obviously, the more elaborate any of these features become, the more expensive and complex the mold. The author is not advocating making any snap-fit application more complex than it needs to be, but these kinds of options are available if needed.

Rules for using assist enhancements are:

- Protect the lock feature against over deflection during disassembly, particularly if tools are used or the lock is not visible.
- Indicate operation of the assist with visuals if necessary.
- If tools are required, design the assist so that readily available tools can be used. Screwdrivers, thin blades (as on a knife or paint scraper) and steel rods (nails, paper clips, etc.) are common tools and will generally meet disassembly needs. The access hole shape (acting as a visual) can sometimes indicate the tool required to release the lock.

4.3.3 User Feel

User feel is strongly related to the same concepts as operator feedback. The kind of tactile and audible signals that can make assembly easier for the operator can also improve the customer's perception of quality in a moveable application. (Recall that *moveable* was defined as one of the application functions.) Movement can be either free or controlled. In a controlled movement application, the customer will be a "user" of the snap-fit. User feel is also more significant in applications used frequently and involving higher forces. For example, user feel in a battery access panel on a TV remote control is much less important than it is for a frequently used appliance cover.

The concepts of the assembly and separation force-deflection signatures also apply to user feel. A solid and firm feeling of engagement accompanied by a smooth, over-center feel for both assembly and disassembly will give an impression of quality.

A good application example in an automobile is a center armrest cover that opens to a storage compartment. Sometimes the latching mechanism is a snap-fit. A console door gets a lot of use; every time it is opened and closed, it can be a reminder of "quality", for better or worse. It is a simple matter to design the lock feature to give good feedback to the customer, and it is free.

If the application uses a releasing lock, you must also pay attention to separation-feel. Design both the insertion and retention faces to give high-quality tactile feedback to the user. Design moveable snap-fits to close with a solid and reassuring sound like a "thud" or "thunk" rather than a cheap sounding "click". Obviously this is somewhat subjective but, for the consumer, it is easy to recognize.

As a side benefit to improved assembly feel when a contoured insertion face profile is used, the assembly force is reduced. This means the stresses on the lock pair are reduced. In a frequently used application, this can prevent long-term lock feature failure.

4.4 Enhancements for Snap-Fit Performance

Performance enhancements ensure the snap-fit attachment performs as expected. While we can perform feature analysis and other evaluations to ensure strength and reliability, sometimes product design parameters such as material requirements or wall thickness can severely limit a locking feature's retention strength. No matter what we do to the lock itself, we simply cannot make it strong enough. Other times we would like to prevent damage to a lock feature or provide insurance that a costly part is not ruined if a lock feature breaks. Performance enhancements include:

- Guards to protect sensitive lock features from damage.
- Retainers to provide local strength and improve lock performance.
- Compliance provided by attributes and features that take up tolerance and help maintain a close fit between mating parts without violating constraint requirements.
- Back-up locks which provide a second means of attachment if the lock feature should fail to work or suffer damage.

4.4.1 Guards

Guards, Fig. 4.12, protect other (weaker) features. Because some locking features are flexible and usually weak in bending, guards are used when it is necessary to protect the

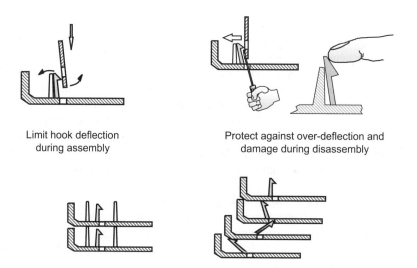

Limit hook deflection during assembly

Protect against over-deflection and damage during disassembly

Protection against stacking, shipping and handling damage

Figure 4.12 Guards protect relatively weak features from damage

lock. Conditions that create the need for guards should generally be avoided, but design constraints may force those conditions. Because the cantilever hook type of locking feature is most likely to require guards, the example shown here involves cantilever hooks. The principles behind the use of guards, however, apply to any locking feature or other sensitive part feature. A number of situations may call for guard features.

Snap-fit features (or other part features) may be in exposed locations and susceptible to possible damage when parts are stacked for shipping and handled before or after shipping. They may be subjected to short or long term deflections. Guards can provide protection and prevent permanent set or breakage. For efficient design, the guard function, if needed, should be built into guides or locators.

When (non-releasing) locks must be manually deflected to release parts, the possibility of over-deflection exists. When the lock is hidden and a tool must be used instead of a finger, the chances for damage increase. Because plastic performance is very time dependent, a lock that survives a very short term deflection during assembly without damage may not survive a similar deflection of longer duration during much slower manual disassembly. Guards can limit lock deflection to just that needed for release and prevent permanent damage.

Sometimes during assembly, a hook can be deflected beyond a safe strain level. In a manner similar to that for preventing over-deflection during disassembly, a guard can limit hook assembly deflection by effectively increasing the hook's bending spring rate. The increased hook stiffness transfers some deflection to the mating part. This will come at a cost, however, because a higher assembly force is now required to deflect the mating part.

4.5.2 Retainers

Preferred practice is to design the attachment's strength and retention performance directly into the locking features. However, design constraints, material requirements or compliance in the parts themselves may result in an inherently weak lock. Sometimes, locks must resist high removal forces and this capability cannot be guaranteed through the lock design alone. Retainers can improve a lock's retention strength by increasing its bending spring rate, or by providing positive interference against deflection, Fig. 4.13. It is appropriate to use retainers for improving both releasing and non-releasing locks. Even non-releasing locks can release under very high load conditions due to gross distortion of the part or the lock itself. A retainer can be positioned to prevent that gross distortion.

Lock features mounted on weak, flexible walls will have limited strength. Retainer enhancements can add local strength within the lock pair.

4.4.3 Compliance

Compliance is the attachment's ability to accommodate dimensional variation so parts are easy to assemble while maintaining a close fit with no looseness. Compliance is also discussed in Chapter 5.

116   Enhancements   [Refs. on p. 134]

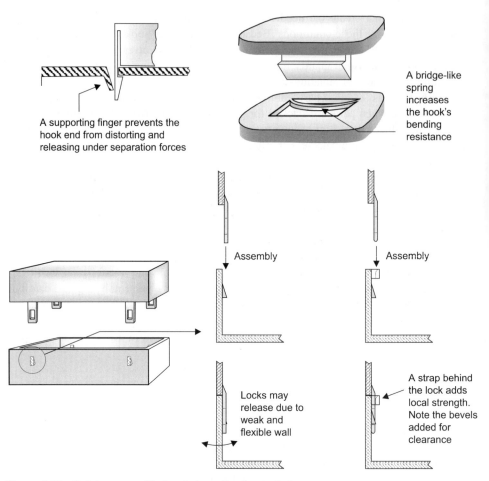

Figure 4.13   Retainers provide local strengthening to locks

Robustness to dimensional variation is designed into the attachment system through proper constraint and constraint feature selection. Sometimes, however, this is not enough to ensure a close fit between parts. Compliance within a constraint pair is then used to supplement the systems performance.

Because they do not use clamp load (like threaded fasteners), a major design requirement for a snap-fit is that parts fit together tightly, with a line-to-line fit, for functional integrity and appearance. Another benefit when a line-to-line fit is maintained is that noise (generally squeaks and rattles) resulting from transient loads is eliminated. One way to get a line-to-line fit is by specifying very close tolerances on parts, but this can be expensive.

Noise in an interface results from energy inputs that cause parts to separate and snap back, resonate or rub together. Preventing noise is a matter of holding parts tightly so separation and relative motion cannot occur under high transient loads or high frequency vibration conditions. The kinds of load cycles that cause noise in plastic are generally of very short duration whereas published performance data is usually based on loads applied over a longer period. Because of the time-dependent behavior of polymers, the strains in the plastic parts under these loading conditions may sometimes exceed the tested strain limits of the material for loads of longer duration. In any case, the compliance features must be designed for long-term effectiveness and resistance to cumulative damage over time. Long-term plastic creep and degradation of the material's properties must also be considered.

When considering how to effectively add compliance to the application, know the significant tolerances and stack-ups in the interface. Whenever possible, design so that tolerances can be taken up in a non-critical area and direction. Know where potential looseness or interference may occur due to differences in the mating materials' coefficients of thermal expansion. Design so that looseness will not cause noise and interference will not cause yield to the extent that looseness results.

See the discussion of critical directions and tolerances in Chapter 5. Deciding where to add compliance depends on critical alignment requirements and interface forces. In general, compliance is added *opposite* the fine-tuning enhancements.

Two ways to add compliance in the interface are *local yield* and *elasticity*. A third way involves adding additional pieces called *isolators* to the system.

4.4.3.1 Compliance through Local Yield

Local yield involves using features within a locator pair to create low levels of interference (through compressive stress), Fig. 4.14a. The interference results in (local) yield within that locator pair. While creep to a lower stress may occur over time because of the compressive stress, as long as no significant additional loads or deflections are applied to cause further yield, a line-to-line fit will be maintained.

Designing for local yield may conflict with the need for good tactile assembly feedback and caution is required so feedback quality is not compromised. Local yield means that resistance to assembly forces will occur over a longer period and involve more than just the lock features. Treat it as an offset or a zero-shift in the assembly force-deflection signature. Sometimes you must balance the two requirements, realizing there is not a perfect solution.

Compliance through local yield should occur within a constraint pair. Compliance between constraint pairs across different locating sites should generally be avoided because it can violate the rule against over-constraint. However, it is sometimes necessary in applications where the base part is an opening or a cavity. An example is the solid to opening application shown in Fig. 4.4. In that application, darts could be added to the land locators that are visible in the illustration. The darts would act against the land-edge locator pairs on the hidden sides of the solid.

While plastic yield is possible in tension, bending or compression, the only mode of yield recommended for yield compliance is compression.

Because local yield requires strength in the features to force the compressive stress, it is rarely employed with locking features. Most of the time, local yield compliance will be found on locator features. Methods of obtaining interference through local yield include *darts*, *crush ribs*, and *tapered* features, Fig. 4.14a.

Darts on pins, lugs and wedges will embed into the edges of other locator features as the parts are pushed together. Darts, to be effective, should be placed on the harder of the two plastics in the interface. When the plastics are similar in hardness, a shallower included angle on the dart can ensure its effectiveness.

(a) Getting compliance through local yield

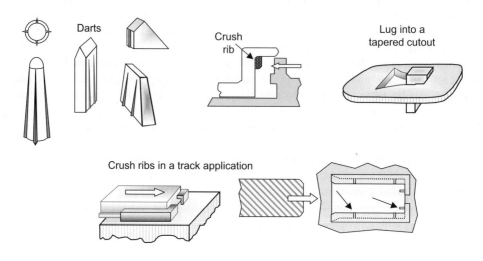

(b) Getting compliance through elasticity by designing spring features into the system

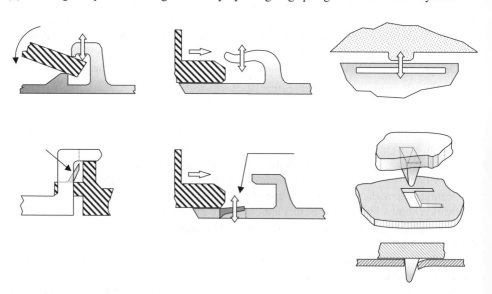

Figure 4.14 Compliance helps take up tolerances

Crush ribs are exactly what the name implies. They are relatively thin ribs that are literally crushed or bent out of the way by the mating feature. The portion of the rib that remains then fills the gap between the parts. In the track application, Fig. 4.14a, strategically placed crush ribs ensure that a bayonet type mount will remain tight in the mating track [4].

Another way to get local yield is by using lugs with tapered cutouts, tapered pins with holes and tapered wedges with slots, Fig. 4.14a.

4.4.3.2 Compliance through Elasticity

The inherent elasticity of plastic can also be used to establish and maintain a line-to-line fit between parts. If parts are structurally rigid, special molded-in features acting as springs can provide elasticity, Fig. 4.14b.

The slight warping that occurs in some parts, particularly panels, as they come out of the mold may provide sufficient residual elasticity for a close fit after the part is nested and locked in place against the base part.

Unlike local yield, which is best limited to locators, elasticity can be effectively used with both locator and lock features. Remember, however, that (most) lock features are weak. Use caution when taking up compliance in a lock pair.

4.4.3.3 Isolators

As a last resort, because it will add cost to the attachment, isolating materials can be added to the attachment to force a line-to-line fit. These can take many forms, including adhesive-backed foam or felt products, soft rubber or felt washers and O-rings. Off-the-shelf O-rings can be easily slipped over a protrusion feature for a quick fix to a looseness problem. (Ensure there is no oil on the rings that could react with the plastic material and that the plastic is not reactive to the O-ring material itself.) Also, ensure that any added materials do not create excessive stress or strain in lock or locator features.

4.4.4 Back-Up Locks

The last enhancement for performance, a back-up lock, provides a locking alternative in the event the intended integral lock feature cannot provide reliable locking. Usually they are simply provisions in the mating and base parts for threaded fasteners, push-in fasteners or metal clips should the integral lock feature fail.

The cost saving potential of snap-fits often indicates their suitability as the mainstream attachment design for an application. However, technical and/or business issues may prevent their serious consideration. Back-up locks can help overcome some of those obstacles. When appropriate, a back-up lock can be made a part of the business case when evaluating an application's attachment alternatives. A significant factor in back-up lock decisions is the piece cost of the part in question. A back-up lock may not be cost effective on a small inexpensive part but could be very desirable on a large, complex and expensive one.

In some applications, a conservative approach to the snap-fit is desirable because the snap-fit may represent a technical "reach". The potential benefits may be substantial but the risk of committing to a snap-fit may preclude its consideration unless a back-up fastening method can be designed into the interface at the same time. Once the design is proven in testing and production, the back-up lock can be eliminated. Should the snap-fit prove unreliable, the back-up lock allows the development program to continue with a reliable attachment for that application.

While the design itself may not be a technical reach, incomplete data about the service loads, material properties or other application requirements may add uncertainty to the design. A back-up lock can allow the snap-fit design to proceed with the confidence that a reliable attachment is possible if the snap-fit does not work.

The design may be such that the locking features of the snap-fit are susceptible to bending or breakage during shipping, handling, assembly or disassembly. If the features cannot be protected by design (see guards) and damage that would render the lock unworkable is possible, a back-up lock ensures the entire part will not be lost because of damage to one feature.

If parts are intended for new designs and also expected to be used on existing designs without provisions for snap-fits, allowing for both methods of attachment accommodates both applications without creating a second set of parts.

Any fastening method may be a candidate as a back-up to a snap-fit and the design criteria should be appropriate to the technology. The same reliability considerations must be applied to the back-up lock as to the original snap-fit.

Back-up locks need not be complex. Providing several clearance holes in a part and pilot holes, bosses or clearance holes in the mating part may be sufficient. Of course, if the back-up lock may become the mainstream design for production then all assembly and processing considerations must be included in the design. If necessary, clearance holes for threaded fasteners can be skinned over and drilled out if needed. Complementary ribs can be added on both parts in proper positions to accept and engage spring steel clips as back-up fasteners.

When a back-up lock is specified because of possible damage in disassembly for service or as a second attachment method on a service part, original assembly issues are no longer critical. Give consideration instead to the tools and fastening methods required for service by the customer or service technician. Do not design a back-up lock that requires special fasteners or special tools.

Rules for back-up locks include:

- Use fasteners identical to other fasteners in the product.
- Use common fasteners that repair facilities are likely to have.
- If high strength is not an issue, and it usually is not in a snap-fit application, design for hardware store type fasteners readily available to the home mechanic.
- Provide adaptable interfaces that permit several sizes, styles or lengths of screw.

4.5 Enhancements for Snap-Fit Manufacturing

Manufacturing enhancements are techniques that support part and mold development, manufacturing and part consistency. Many are documented in standard design and manufacturing practices for injection-molded parts and are already recognized as important factors in plastic part design. They fit neatly into the Attachment Level Construct as enhancements.

These enhancements generally make the part easier to manufacture. Parts that are easier to make are more likely to be made consistently and correctly. They are more likely to perform as expected, an important component of reliability. Another benefit is that they are likely to be less expensive.

Manufacturing enhancements can provide benefits in:

Cost	Shape consistency
Appearance	Mold development
Reliability	Internal stresses
Process cycle time	Performance consistency
Fine-tuning for development	Adjustments for variation

Detailed plastic part design principles, mold design practices and manufacturing procedures are well documented in many other books and standards and that information will not be repeated here. This section is not intended to be a comprehensive guide to the subject of mold design. The intention is to simply capture this particular aspect of snap-fit design as an enhancement and present a few of the more basic concepts that relate directly to snap-fits.

Remember that snap-fit features are subject to the same rules of good mold design as the other features in an injection-molded part. Many snap-fit features are protrusions from a wall or surface and they should be designed according to the same rules as protrusions.

Sometimes, a snap-fit designer relies on the part supplier (if another company) or the experts in their own company to provide the information and design expertise for part processing. There is nothing wrong with this; one should rely on the experts. However, it does not hurt to know enough to be able to ask some intelligent questions. You may occasionally catch something they have overlooked. The part designer is also most familiar with the requirements of the application and is in the best position to ensure they are properly considered.

Manufacturing enhancements fall into two groups. Those that improve the part making process we call *process-friendly*. Those that allow for relatively easy dimensional changes to the mold, are *fine-tuning* enhancements.

4.5.1 Process-Friendly

Process-friendly design is simply following the recommended and preferred plastic part design practices. Process-friendly parts are robust to the molding process and are likely to be higher quality, less expensive and more consistent in performance than parts that are not.

The information shown in this section was drawn from a number of publications. It seems to represent general design knowledge because very similar or identical information was typically found in multiple documents. Rather than cite numerous publications for each item presented all the publications are listed at the end of this chapter.

The single most important rule is to keep the design simple: the simplest design that will work is obviously the best, Fig. 4.15a. Simple feature designs mean less costly molds and greater consistency. When moving parts are required in the mold to make under-cuts and hidden features, die complexity and cost goes up. Access for molding under-cuts is an ever-present issue with mold design and snap-fits are no exception. Features that can be produced without requiring the added complexity of mold features like slides and lifters are always preferred.

(a) Use simple shapes and allow for die access and part removal

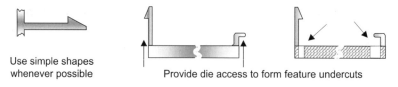

Use simple shapes whenever possible

Provide die access to form feature undercuts

(b) Round all corners, both internal and external

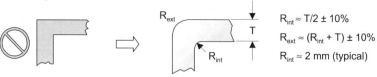

$R_{int} \approx T/2 \pm 10\%$

$R_{ext} \approx (R_{int} + T) \pm 10\%$

$R_{int} \approx 2$ mm (typical)

(c) Adjust the protrusion thickness relative to the wall thickness and use a radius at the wall

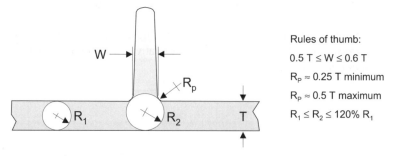

Rules of thumb:

$0.5\,T \leq W \leq 0.6\,T$

$R_p \approx 0.25\,T$ minimum

$R_p \approx 0.5\,T$ maximum

$R_1 \leq R_2 \leq 120\%\,R_1$

1. Calculate the basic protrusion width (W) from the wall thickness.
2. Add the draft angle to the basic protrusion width.
3. Add a radius (R_p) at the protrusion base.
4. Verify that the material volume at the protrusion base does not exceed about 120% of the normal wall volume.

Figure 4.15 Common process-friendly design practices

4.5 Enhancements for Snap-Fit Manufacturing   123

(d) Protrusion spacing

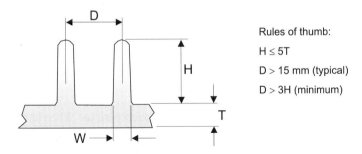

Rules of thumb:

H ≤ 5T

D > 15 mm (typical)

D > 3H (minimum)

(e) Allow for draft angles

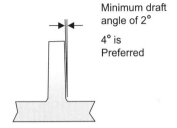

Minimum draft angle of 2°

4° is Preferred

(f) Taper all section changes

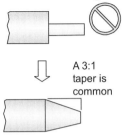

A 3:1 taper is common

(g) No thick sections

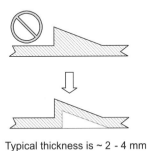

Typical thickness is ~ 2 - 4 mm

(h) Allow for a shut-off angle where the die faces meet in shear

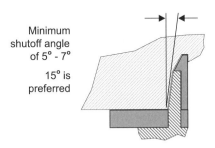

Minimum shutoff angle of 5° - 7°

15° is preferred

Figure 4.15 (*continued*)   Common process-friendly design practices

Sometimes, a complex feature shape may be required if moving parts in the die are to be avoided. In that case, consider the costs and advantages of both designs. Also, consider that analytical tools for predicting lock and locator behavior tend to be less accurate as feature shapes become more complex.

Specify a radius for all inside and outside corners, Fig. 4.15b. The idea is to avoid all sharp corners and maintain a constant wall thickness for smooth plastic flow through the mold. (The melt front does not like surprises.) Corners cause turbulence and are hard to fill. It is not enough to simply ask for fillets and radii in a general drawing note. Put a dimension at every site where a fillet or radius is required.

Sharp internal corners also create sites for stress concentrations. When at the base of a constraint feature, they can cause feature failure. Treat every protrusion feature (hooks, pins, tabs, lugs, etc.) as a rib and follow the guidelines for rib sections and rib spacing. The idea is to maintain a relationship between the wall thickness and the protrusion thickness so that voids or residual stresses at the base of the feature do not occur. Some basic rules are shown in Fig. 4.15c and Fig. 4.15d. Keep in mind however that these are general rules and simply provide a good starting point. Specific plastics can have their own requirements.

If a prototype part shows sink marks on the opposite side of the wall from a protrusion, this is a good indication that voids or residual internal stresses may be present at the base of the feature. These will weaken the feature and may result in failure.

Include a draft angle. This allows the part to be easily removed from the mold. Start with the basic feature size then add the angle to each side, Fig. 4.15e.

Avoid thick sections and abrupt section changes for the same reasons you avoid sharp corners. Another reason is the difficulty of cooling a thick section of plastic. To properly cool a thick section results in significantly longer cycle times and higher cost, Figs. 4.15f and g.

Where die faces come together in shear, a shut-off angle is necessary, Fig. 4.15h. This applies when access for molding hooks or lugs is required, Fig.4.15a.

Gates are the areas where the plastic melt enters the mold cavity and gate style and location are other aspects of mold design that can have a significant effect on the snap-fit features. Gates can affect the constraint feature's location (due to part warping) and the feature's strength. Remember that the mold designer is not likely to know the critical areas of your design and will put the gates at locations they believe are the best sites for mold fabrication and performance unless you indicate otherwise. Gates should be located:

- Away from flexible features and impact areas.
- So that knit lines will not occur at high stress areas, including living hinges.
- In the heaviest/thickest sections so that flow is to the thinner, smaller areas.
- So flow is across (not parallel to) living hinges.
- So flow is directed toward a vent.
- In non-visible areas.
- So that flow distance to critical features is not excessive.

Gate location can also affect part warpage. Be sure the snap-fit features do not move out of position due to excessive part warpage. If they do, guide enhancements may be needed to bring the locks back into proper position for engagement.

4.5.2 Fine-Tuning

Fine-tuning involves adjusting the mold dimensions to result in correct final part dimensions. It is necessary because the nature of the molding process is such that first parts out of the mold will not be perfect. Despite the use of predictive tools and highly controlled processing techniques, one never knows exactly what the part will be like until first parts are made. This is particularly true when the snap-fit designer is concerned with high precision in constraint feature locations and dimensions. Part changes and adjustments during part development become much easier when allowances are made for fine-tuning during part design.

Once production begins, long-term wear, variations in raw materials, design changes and variation in the other part may also require periodic mold adjustments to maintain attachment quality throughout the part's production run.

In anticipation of changes, plan for easy mold adjustments at strategic locations. The purpose is to avoid large-scale (expensive and time-consuming) mold changes. In other words, make the snap-fit interface "change-friendly".

The first step in adding fine-tuning enhancements is to identify critical alignment and load carrying requirements and the constraint sites that provide that capability. This should have already occurred in the design process because you needed to understand the critical constraint sites to establish constraint and compliance requirements. These sites represent the areas of the part (thus the mold) where fine-tuning is likely to be needed, Fig. 4.16. Fine-tuning site selection also affects compliance enhancement locations. Once these critical sites for fine-tuning are identified, you can decide if *metal-safe* design or *adjustable inserts* are appropriate.

Metal-safe means to fine-tune the part by removing rather than adding metal to the mold. Obviously, it is much easier to simply grind material away in the mold than to first build up an area then shape it by grinding metal away. Once the critical sites have been identified, select initial nominal dimensions and tolerances at or slightly beyond the minimum material condition, Fig. 4.17. Be careful not to carry the idea of metal-safe design to such an extreme

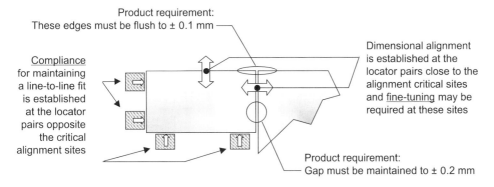

Figure 4.16 Selecting sites for compliance and fine-tuning

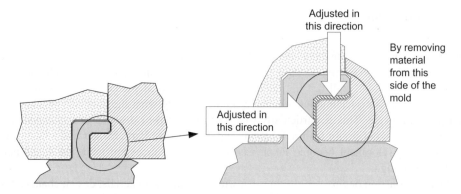

Figure 4.17 Metal-safe fine-tuning on a lug

that first parts out of the mold are not even close to design intent. This will render the parts useless for fine-tuning and just add more work.

Adjustable inserts can also be used to permit fine-tuning critical dimensions on constraint features. Inserts are easily removed from the mold and can be modified and reused or replaced by other inserts, Figs. 4.18 and 4.19. Unlike metal-safe design, inserts allow critical dimensions to be easily adjusted in both directions, either adding or removing material.

(a) Panel to cavity application

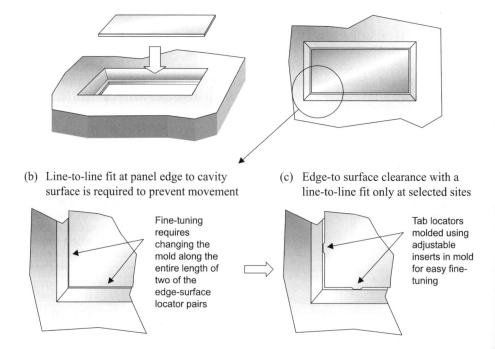

Figure 4.18 Fine-tuning with adjustable inserts

(a) Initial design leaves some clearance at the hook

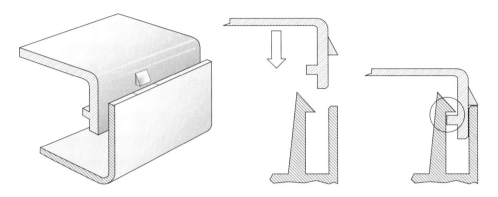

(b) Fine-tuning at the edge using an adjustable insert brings the hook face into line-to-line contact with the mating surface

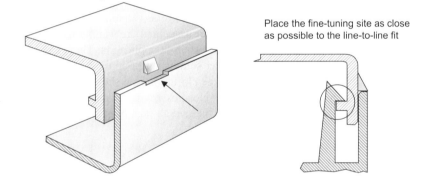

Place the fine-tuning site as close as possible to the line-to-line fit

Figure 4.19   Fine-tuning with adjustable inserts

Use of adjustable inserts requires designing for *local adjustment* at the critical constraint sites. This means you have provided distinct locator features in those areas rather than using a large part area such as a surface or edge as a natural locator. Fine-tuning a locator feature or features is much easier than changing the mold for a major part feature.

In the application shown in Fig. 4.18, rather than locate at the edge to surface interface (natural locators) the fit of the panel to surface is controlled at specific contact sites around the part perimeters. Fine-tuning adjustments can be made by modifying the inserts at these sites rather than changing the entire part.

Some rules for fine-tuning are:

- Identify the constraint sites that provide critical positioning or alignment. Make allowance for fine-tuning at these sites.
- Identify the constraint features that provide the critical strength in the attachment and determine if fine-tuning will be necessary to adjust performance. Keep in mind that

simply increasing strength by adding thickness is limited by the process-friendly rules. Strength can also be increased by adding structural ribs to the features. These ribs can also be fine-tuned for performance.
- In general, compliance enhancements should be placed at locator pairs that are not fine-tuning sites.
- Select the initial nominal dimensions and tolerances between those sites so that the minimum material condition will occur at the tolerance range maximum. This will put the features slightly undersize.

4.6 Summary

This chapter provided detailed descriptions of enhancements and rules for their usage. Enhancement features are one of the two physical elements of a snap-fit. They may be distinct physical features of an interface or attributes of other interface features. Enhancements improve the snap-fit's robustness to the variables and unknown conditions that can exist in manufacturing, assembly and usage and are summarized in Table 4.2.

Enhancements are often subtle details in a snap-fit application. They may not be obvious at first glance. It is suggested that the reader study snap-fit applications to become familiar with the usage of enhancements. The luggage closure buckle shown in Fig. 4.20 is a readily available application. If you can compare closures from several manufacturers, you will begin to see how enhancements can affect the overall quality of the application.

4.6.1 Important Points in Chapter 4

Some enhancements are required in every application; others depend on specific needs of the application, Table 4.3. When soliciting bids on a snap-fit application, the required enhancements should be made part of the business case and considered non-negotiable. They are almost as essential to ensuring a high quality and successful snap-fit as are the constraint features. When bidding on an application, enhancements may be the attention to detail that wins you the contract.

During snap-fit development, include enhancements in the initial attachment concepts and in the first detailed parts made when possible. However, including all enhancements in the original design or even the first prototype parts is usually not possible or practical. One must actually assemble and disassemble actual parts to properly assess the need for some enhancements. Desktop manufacturing methods can provide pre-prototype parts with enough detail that requirements for visuals, guides and assists can be identified. Other enhancements (assembly feedback and user-feel, for example) usually require that parts be made from the design intent plastic using production molds to properly identify and develop enhancement details to meet product requirements. The need for retainers may not be apparent until parts undergo physical testing. Table 4.4 shows the steps in the snap-fit

Table 4.2 Enhancements Summary

Name	Why	What/How	Notes
Ease of assembly			
Guidance	Ease of assembly	Guide—stabilize parts Clearance—no interference Pilot—correct orientation	No simultaneous engagement, use locators. Usually a feature attribute. Use locator or guide if possible.
Feedback	Indicate good assembly	Tactile, audible, visual signals and consistent behavior	Maximize positive signal and minimize system "noise". May conflict with compliance.
Activation and usage			
Visuals	Disassembly and operation information	Text, arrows, symbols	Standard symbols should be used. Training and awareness for customers and service are needed.
Assists	Enable disassembly or operation	Extensions for fingers, tool access	For non-releasing locks. Possible feature damage, visuals or guards may be needed
User feel	Perceived quality	Tactile, audible	Manually activated locks in moveable applications.
Performance and strength			
Guards	Protect weak or sensitive features	Prevent over-deflection, reduce strain	Cantilever hooks in particular may need protection.
Retainers	Improve retention performance	Strengthen or support the lock or stiffen the lock area	Cantilever hooks in particular, may sometimes need retainers.
Compliance	Take up tolerances and prevent looseness	Elastic features or local yield	May interfere with feedback, use care.
Back-up lock	A back-up attaching system	Readily available fasteners, adaptable interfaces	For service and repair or as an alternative mainstream design.
Manufacturing			
Process-friendly	Efficient and consistent manufacturing process	Feature design and orientation	Follow mold and product design guidelines.
Fine-tuning	Development and manufacturing adjustments	Metal-safe design, adjustable inserts, local adjustments	Don't over-do metal-safe. Select sites carefully. Use at locator sites controlling critical dimensions.

(a) For assembly

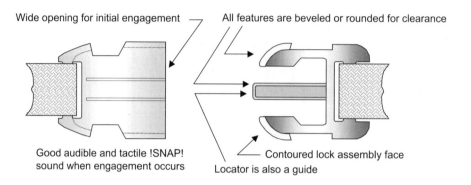

(b) For disassembly

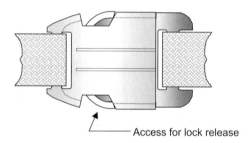

Figure 4.20 Enhancements in a luggage closure, a common application

development process where one is most likely to have enough information to add a particular enhancement. Of course, enhancements may also be added after the fact in response to product problems.

Some designers seem to feel as if they have somehow failed in their snap-fit design if they must add various enhancement features. That impression is the result of applying traditional threaded fastener thinking to snap-fits. Remember that a threaded fastener attachment represents a "brute strength" approach to fastening. The interface details required for a good snap-fit design go beyond those necessary for a threaded fastener attachment. In reality, enhancements belong in every snap-fit and an application without them will not be the best possible design.

As with the other physical features of snap-fits, locks and locators, the Attachment Level Construct does not pretend to have invented enhancements. Certain enhancements, particularly those related to manufacturing and design for assembly, are well documented elsewhere; others are not. In both cases, however, the construct effectively captures them, providing a means of describing and classifying them for use. Most of the examples of enhancements shown here were found on products, often on many different products in many different variations.

Table 4.3 Enhancement Requirements

Group	Enhancement type	Required in all applications	Required in some applications	Nice to have and recommended
Assembly	Guidance (Guides)	✓		
	Guidance (Pilots)		✓	
	Guidance (Clearance)	✓		
	Operator feedback	✓		
Activation	Visuals		✓	
	Assists		✓	
	User feel		✓	
Performance	Guards		✓	
	Retainers		✓	
	Compliance	✓		
	Back-up lock		✓	
Manufacturing	Process-friendly	✓		
	Fine-tuning			✓

4.6.2 Design Rules Introduced in Chapter 3

Guides:

- Lock features should never be the first features to make contact with the other part.
- Guides must engage before the operator's fingers contact the base part.
- Avoid simultaneous engagement of multiple features. One or two guides (or locators) should engage first to stabilize the mating part to the base part, particularly when the guides are protruding features engaging into holes or slots.
- A "tip" assembly motion is preferred because it forces initial engagement at one end of the part followed by rotation to sequentially engage the remaining features.
- Build the guide and pilot functions into existing constraint features whenever possible.

Clearance:

- Specify a taper or a radius on all corners and edges of the parts proper as well as on all the features.
- Always provide generous clearance for initial engagement.

Table 4.4 Enhancement Identification and the Development Process

Enhancement	Development stage				Comments
	R	C	D	T	
Guidance (Guides)		✓			Required. Combine with locators.
Guidance (Pilots)	✓				Required if a symmetric part can be improperly oriented.
Guidance (Clearance)		✓	✓		Required. Certain clearance features (lands) may be identified early. Details of clearance, bevels, radii added during design.
Operator feedback		✓	✓	✓	Required. Details added in design. May require testing and evaluation.
Visuals	✓		✓		Need may be identified but implementation usually delayed until final parts.
Assists User feel	✓	✓	✓	✓	If a non-releasing lock with limited access. If a user activated lock in moveable application.
Guards		✓			Need may be identified early, part stacking or manual deflection for example.
Retainers		✓	✓	✓	Sometimes predictable based on application concept (constraint features on thin walls). Sometimes identified in analysis or test.
Compliance		✓	✓	✓	Identify sites at concent development. Execute details during detailed design.
Back-up lock	✓		✓	✓	May be early or after testing indicates potential problem.
Process-friendly		✓	✓		Feature orientation decisions during concept. Details and dimensions added during design.
Change-friendly			✓		Details and dimensions added during design.

Development stage symbols:
R—When establishing specific application requirements.
C—While developing the attachment concept.
D—Detailed design and analysis.
T—Testing.
✓—Need for enhancement is likely to be first identified.
✓—Follow-up or secondary identification.

4.6 Summary

Operator feedback:

- Ergonomic Factors—Assembly forces must be within an acceptable range. A comfortable operator position, normal motions and parts that assemble easily will help create a work environment in which the operator can be sensitive to tactile feedback.
- Avoid extreme rotational or reaching motions.
- Avoid high forces on fingers, thumbs or hands to install a part.
- Design for top down, forward and natural motions carried out from a comfortable body position. Avoid awkward reaches or twisting motions and reaches over the head.
- Provide solid pressure points. Weak parts or soft materials may require local strengthening.
- Design to ensure positive and solid contact between strong locator features.
- A rapid lock return can give a good audible and tactile signal that the lock is engaged.
- A strong "over-center" action as a lock engages will give a feeling that the part is being pulled into position.
- Consistency in part assembly performance, through process-friendly design, allows the operator to acquire a feeling for a good attachment.
- Provide highly visible features that are clearly aligned when the assembly is successful.
- Design for go/no-go latching so a part that is not properly locked in place will easily fall out of position to create an obvious assembly failure.

Assists:

- Indicate operation of the assist with visuals if necessary.
- If tools are used or the lock is not visible, use guards to protect the lock feature against over deflection during disassembly.
- If tools are required, design the assist so that readily available tools can be used.

Compliance:

- Compliance through local yield should occur within a constraint pair.
- Yield compliance should only involve plastic yield in compression.

Back-up locks:

- Design to use fasteners like those already present in the product.
- Use common fasteners that repair facilities are likely to have.
- If high strength is not an issue, design for hardware store type fasteners readily available to the home mechanic.
- Provide adaptable interfaces that permit several sizes, styles or lengths of screw.

Process-friendly:

- Refer to the published rules and guidelines for mold design.
- Consider all protrusion features as ribs and follow rules for rib design and spacing.
- Always specify radii and smooth transitions.
- Locate gates away from flexible features and impact areas.
- Locate gates so that knitlines will not occur at high stress areas, including living hinges.
- Locate gates in the heaviest/thickest sections so that flow is to the thinner, smaller areas.
- Locate gates so flow is across (not parallel to) living hinges.

- Locate gates so flow is directed toward a vent.
- Locate gates in non-visible areas.
- Locate gates so that flow distance to critical features is not excessive.

Fine-tuning:

- Make allowance for fine-tuning at constraint sites providing critical positioning or alignment.
- Increasing feature strength by adding thickness is limited by the process-friendly rules. Strength can also be increased by adding structural ribs to the features.
- Use compliance enhancements at locator pairs that are not fine-tuning sites.
- Set nominal dimensions and tolerances at slightly below the minimum material condition for metal-safe design at selected sites.
- Fine-tuning sites should be as close as possible to the critical dimensions that must be controlled.

References

1. From a 1994 conversation with Rich Coppa, Senior Principle Engineer, Camera Division of the Polaroid Corporation, Boston MA.
2. The application redesign in product example #3 was developed by Tom Froling and Tom Nistor.
3. Bonenberger, Paul R., The Role of Enhancement Features in High Quality Integral Attachments (1995), Technical paper #294 at Society of Plastics Engineers' Annual Technical Conference '95, Boston, MA.
4. From a shutter assembly on a Polaroid camera (model unknown).

Bibliography

The following publications all provide highly useful information on plastics and designing for injection molding. They were used as reference for this chapter.

In alphabetical order:

Beall, Glenn L., *Plastic Part Design for Economical Injection Molding*, 1998, Libertyville, IL.
Dupont Polymers, *Dupont Engineering Polymers—Product Information Guide*, Dupont Polymers Department, Wilmington, Delaware.
GE Plastics, *GE Engineering Thermoplastics Injection Molding Processing Guide*, 1998, General Electric Company, Pittsfield, MA.
Hoechst Technical Polymers, *Designing With Plastic—The Fundamentals*, Design Manual TDM-1, 1996, Ticona LLC, Summit, NJ. (Formerly Hoechst Celanese Corporation, now a division of Celanese AG.)
Malloy, Robert A., *Plastic Part Design for Injection Molding—An Introduction*, 1994, Hanser/Gardner Publications, Inc., Cincinnati, Ohio.
Molders Division of The Society of the Plastics Industry, Inc., *Standards and Practices of Plastics Molders—1998 Edition*, Washington DC.
Monsanto Company, *Monsanto Plastics Design Manual*, 1994, Monsanto Company, St. Louis, MO.
Xerox Corporation, Plastic Design Aid (wallchart), 1987.

5 Other Snap-Fit Concepts

Rather than interrupt other topics with a detailed discussion of the concepts of constraint and decoupling, those subjects are covered in detail here.

Constraint was introduced in Chapter 2 as the most fundamental of the four key requirement for snap-fits. It was also discussed in Chapter 3 in terms of the application and use of locators and locks as constraint features.

Decoupling was referred to briefly but has not been discussed at length. Decoupling is the extent to which a locking feature's assembly and retention behaviors are independent of each other. It has important ramifications for understanding lock behavior and improving lock performance.

5.1 The Importance of Constraint

Conscious or explicit consideration of constraint in attachments is not common practice. Many designers are accustomed to specifying threaded fasteners and are familiar with design practices for attachments using threaded fasteners. Threaded attachments achieve constraint in a rather simple manner: fasteners are added and tightened until the resulting clamp load is sufficient to prevent relative motion in the joint. Constraint between the joined parts happens automatically and making explicit decisions about constraint during threaded fastener attachment design is not necessary. As the reader will learn in Chapter 8, Snap-fit Problem Diagnosis, improper constraint is a major contributor to problems with snap-fits.

Some design practices for attachments that use adhesives or other methods that do not rely on clamp load are similar to snap-fit design but not identical. There are special issues with snap-fits that are not present in any other attachment. Designers must always be aware that many design principles associated with other attachment methods do not work for snap-fits.

Most importantly, and unlike threaded fasteners, it is not possible to get tensile stretch in snap-fit features to create significant clamp load. Getting clamp load through feature bending is possible, but not extremely efficient and is not recommended. In any case, because plastics tend to creep under stress, even if some clamp load is designed into a plastic snap-fit, it will eventually relax and the clamp load will be lost. If the features do not break or yield during this process, you are left with a line-to-line fit. The art of good snap-fit design is to simply design that line-to-line fit into the interface at the start. Proper use of constraint makes it possible to balance the attachment's need for strength, assemblability and a line-to-line fit with the realities of part variation and tolerances.

5.1.1 Constraint Review

Recall that the motion of an object in space is described by six translational and six rotational movements for a total of 12 degrees of motion (DOM). In the case of snap-fits, this is how the positional relationship of the mating part to the base part is described. Constraint features are used to restrict motion and systematically remove degrees of motion from the mating part to base part interface. Some people have an intuitive feeling for constraint and apply constraint principles automatically when designing a snap-fit. For others, an understanding of constraint must be developed.

Because constraint features are restraining the mating part to the base part, constraint (i.e. feature line-of-action) vectors are shown in the figures as acting on the mating part to prevent its movement. Because we are designing snap-fits for line-to-line fit, the "force" represented by constraint vectors is a potential resistance to external loads applied to the system. It is not an actual or a constant force exerted by a constraint feature.

5.1.2 Constraint Principles

When considering constraint, it is important to differentiate between *perfect constraint* and *proper constraint*. For learning purposes, we will first introduce and explain constraint in terms of perfect constraint. Under perfect constraint conditions, forces between all constraint pairs are statically determinate. In other words, we can calculate them using principles of mechanics and statics without worrying about part spring rates or redundant forces. For most applications, achieving perfect constraint while avoiding possible looseness between the parts would require zero tolerances. Zero tolerance is, of course, an expensive and generally impractical situation. Complex part geometry and part compliance also make perfect constraint (easy to get with the rectangular solid used in the following example) rather impractical in reality.

5.1.2.1 Perfect Constraint

Perfect constraint implies perfect or 100% attachment efficiency where part movement is prevented using the minimum number of restraint points and the interface system is statically determinate. An understanding of some of the characteristics of perfect constraint will provide a basis for the more practical concept of proper constraint.

Recall that a plane is defined by three points and a line by two and that a system of three, two and one point(s) can perfectly locate an object. An object in space (the mating part), Fig. 5.1a can be prevented from moving in one DOM by constraining it at three points (a plane) as shown in Fig 5.1b. Next, adding two points to one side of the object will prevent movement in another DOM, Fig. 5.1c. A single point on another side of the object will prevent movement in a third DOM, Fig 5.1d. The object's position is now accurately determined by the plane, the line and the single point. This is acceptable as long as no forces act on the object to move it out of this position. In products, forces are part of the design reality so more than just well-defined positioning is required to constrain the object.

5.1 The Importance of Constraint   137

(a) A rectangular object is to be positioned to another object

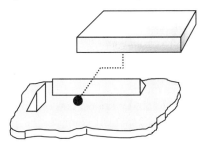

(b) First, three points define a plane

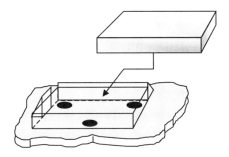

(c) Second, two points define a line

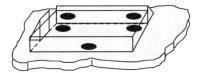

(d) Third, a single point completes the positioning

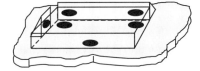

(e) Restraining forces hold the part in position

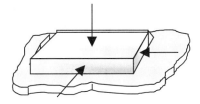

(f) Restraining forces can be composed into one resultant force

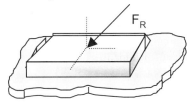

(g) In a snap-fit, line-to-line contact with features holds the part in position

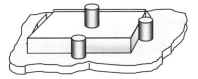

Figure 5.1   Perfect constraint

Additional restraint is needed to hold the object against the plane, line and point. This is done by adding three forces, each one acting to hold the object against one of the three positioning sites as shown in Fig. 5.1e. These three additional forces accomplish a number of important things as they hold the object against the established location points: (1) they prevent translation movement away from the established points and (2) they remove all of the rotational movements from the system. The remaining nine DOM are removed (three in translation and six in rotation) and the object is now constrained in a total of 12 degrees of motion. These three restraining forces can be shown as one resultant force F_R as shown in Fig. 5.1f. This force must be strong enough to hold the object against any outside forces seeking to move it out of position. A bolt passing through the part along the F_R line-of-action and tightened to create clamp load would do this. In the case of a snap-fit, however, we will not rely on clamp load. Features that will not exert clamp load, but will resist movement can be strategically placed so they just touch the object (a line-to-line fit), Fig 5.1g. This represents a perfectly constrained snap-fit. (As with the forces above, the restraining effect of the three features could also be represented as a resultant.)

Recall the discussion in Chapter 3 about the desirability of spacing constraint features as far apart as possible for dimensional robustness and strength. That design rule can be further elaborated using this example of perfect constraint. To maximize mechanical advantage for strength and minimize dimensional sensitivity in each direction, the three planar constraint points should be arrayed against the largest area of the object, the two linear constraint points are arranged against the next largest area of the object and the single point against the third largest area. Compare the inherent stability of the arrangement in Fig. 5.2a to the instability of the arrangement in Fig. 5.2b. While the latter is technically correct with respect to perfect constraint, it obviously lacks the mechanical advantage against outside forces and the dimensional robustness of the former. What if the object is a cube and all sides are equal in size? Some judgement is required depending on the application requirements but, as a rule, the three-point constraint site would be selected to resist the highest forces or to control the most critical dimensions.

(a) Perfectly constrained, robust for locating and stable against outside forces

(b) Perfectly constrained but robustness and stability are poor

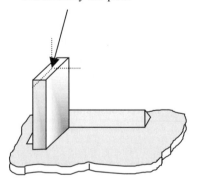

Figure 5.2 Part stability, dimensional robustness and constraint feature strength is optimized by proper feature placement with respect to part shape

The spacing principles for maximizing the object's stability with respect to the initial constraint points are also true for the restraining points that are added to hold the object against the constraint points.

Note that the "theoretical" three-point site may, in reality, not look like three points at all and it is not necessarily the first site of contact between the mating and base parts nor is it necessarily the most constraining locator site.

5.1.2.2 Proper Constraint

Perfect constraint is an ideal. In reality, snap-fit design is a compromise between perfect constraint and the realities of a given application. When we have designed according to the constraint guidelines, we can say the snap-fit is *properly* constrained, meaning that within the limits of tolerances and with the help of local compliance, the attachment is a reasonable approximation of perfect constraint. A realistic explanation of proper constraint is that it exists when there are no *gross* violations of the rules defining improper constraint. It is the absence of *under-constraint* and the minimization of *over-constraint* conditions.

When parts are properly constrained, they will have these desirable characteristics:

- Can be assembled without forcing parts together.
- Normal or loose tolerances between constraint features in the interface are possible.
- Static analysis of forces on the constraint features is possible.
- No residual forces exist between constraint pairs after assembly.

5.1.2.3 Proper Constraint in Less than 12 DOM

We have commonly used a fixed application as an example of proper constraint and required that the mating part be restrained in exactly 12 degrees-of-motion. The reader must not forget that when the attachment's action is moveable (either controlled or free), proper constraint may exist with less than 12 DOM.

5.1.2.4 Under-Constraint

In a fixed application, if parts are constrained in less than 12 DOM, they are under-constrained. Under-constraint can cause the following problems:

- Lock feature damage because locks are improperly loaded.
- Parts improperly aligned or loose because the constraint features cannot effectively prevent relative movement.
- Parts falling off when damaged constraint features release or break.

A common under-constraint mistake is designing so that a lock must carry forces in an improper direction. Locks are weak and should be used only to resist movement in the separation direction. Locator features must be used to prevent all other movements.

A second common under-constraint mistake is failure to place locators for maximum mechanical advantage. This relates to the discussion of stability and Fig. 5.2. Because of highly complex part shapes this becomes a highly subjective area. A locator arrangement may not be technically under-constrained but it may be less stable than it could be. The difference between proper constraint and over or under-constraint is often a matter of degree, not absolutes.

The most important thing to know about under-constraint is that it must be fixed.

5.1.2.5 Over-Constraint

When constraint features are "fighting" each other, they are over-constrained. Over-constraint can cause these problems:

- Difficult assembly. When locator pairs must be forced together, high assembly forces result and immediate damage to constraint features is possible.
- Increased feature stress. Assembly interference between constraint pairs can create internal residual forces. Short or long-term feature damage and failure are possible.
- Part buckling and temperature distortion as joined parts of unlike materials expand and contract at different rates. This is unsightly and may also result in long-term feature damage and failure.

A common mistake is to try fixing an over-constrained design by specifying extremely close tolerances. This will increases the cost of the parts and, while it may eliminate difficult assembly or prevent feature damage during assembly, it cannot fix the thermal expansion/contraction problem. Another common mistake is to confuse over-constraint with higher strength in the attachment.

There are two kinds of over-constraint violations: *features in opposition* and *redundant features*. Features in opposition is the more serious of the two.

a. Over-Constraint Due to Redundant Features

Sometimes, the designer feels compelled to increase strength by adding additional constraint pairs to resist a force. When two or more co-linear lines-of-action are resisting the same translational force, those constraint pairs are redundant in that direction, Fig. 5.3a. In other words, one of the constraint pairs could be removed or changed to eliminate a redundant line-of-action without changing the system's overall constraint condition. That is exactly what you should do. Determine which constraint pair is least effective or more expensive to mold and eliminate it or modify it. Design all the necessary strength into the remaining constraint pair, Fig. 5.3b.

Redundancy in constraint leads to extra cost in the parts because it involves extra constraint pairs and it requires closer tolerances to ensure simultaneous contact of the redundant pairs. Most of the time, however, over-constraint due to redundant features is not serious in terms of attachment performance. With redundant features, we can think of one constraint pair as helping the other (even if that help is unwanted). This is not the case with opposing features.

b. Over-Constraint Due to Features in Opposition

Opposition occurs when two constraint pairs have co-linear lines-of-action that are acting in opposite directions, Fig. 5.4a. Because they have opposing strength vectors, the constraint pairs will fight each other and the potential for damage is high. Unless the tolerances between these pairs are held quite close (on both parts), the chances are good that in most assemblies, there will either be some initial looseness along that axis or the parts will require

(a) Solid to surface application

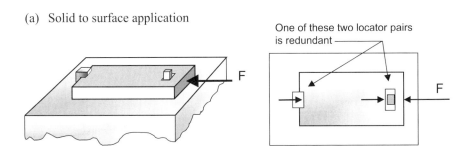

(b) Redundant constraint eliminated

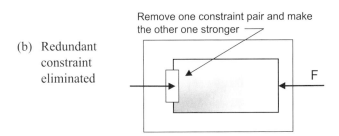

Figure 5.3 Over-constraint due to feature redundancy

additional force to engage because of interference between the pairs. The resulting strain and residual stress can cause the features to relax and loosen over time.

Even if one is willing to pay the price for very close tolerances to prevent looseness or strain between the pairs, Fig. 5.4b, the attachment will not be robust to thermal expansion or contraction along that axis, Fig. 5.4c. If the parts are made of similar materials, the thermal effects may be minimal. However, some plastics can have quite different thermal expansion rates depending on fiber alignment or flow characteristics so having identical materials may be no guarantee against problems. If thermal expansion or contraction is an issue, and features must oppose each other, try to place them as close to each other as possible to minimize the actual size differential when expansion and contraction occurs.

The best fix for features in opposition is to replace or redesign the problem constraint pairs so that motion in both of the directions along the axis in question is resisted at only one of the pairs, Fig. 5.5a. The second choice is to add compliance enhancements at one of the problem sites, Fig. 5.5b. However, if forces are resisted or critical dimensions are controlled by a constraint pair, adding compliance at that site may not be possible.

5.1.2.6 General Constraint Rules

Most of the design rules related to constraint can be found in Chapter 3. Only a few pertinent constraint rules are repeated here as reminders for the constraint worksheet discussion that follows.

(a) Solid to surface application

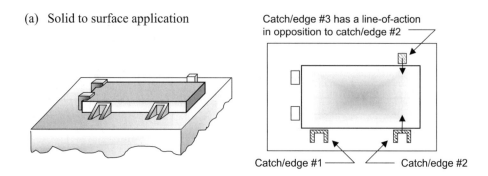

(b) Close tolerances can allow for assembly and no looseness

(c) Close tolerances will not compensate for thermal effects

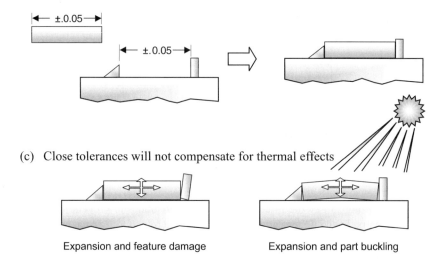

Figure 5.4 Features in opposition

- Fixed snap-fits are properly constrained in 12 DOM
- Moveable snap-fits can be properly constrained in less than 12 DOM
- Locator features are strong so use them to remove as many DOM as possible. Minimize the DOM removed by (weak) lock features.
- The *tip*, *slide*, *twist* and *pivot* assembly motions tend to maximize DOM removed by locators and are preferred for strength. The *push* assembly motion generally maximizes DOM removed by locks and is not preferred.
- Over-constraint due to opposing constraint pairs is undesirable and it should be fixed if possible. Sometimes, however, it is a practical necessity. To compensate, use compliance enhancements or if thermal effects are minimal, close tolerances between the constraint pairs can be used.
- Over-constraint due to redundant constraint pairs is inefficient and unnecessary.
- An under-constraint condition is unacceptable and must be fixed.

(a) If forces exist in both directions, redesign to restrain movement at one constraint pair

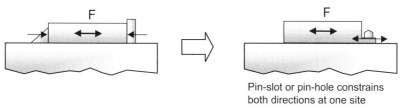

Pin-slot or pin-hole constrains both directions at one site

(b) If force or alignment requirements are in only one direction, compliance can be used

If there is an external force acting at only one site, add compliance at the other

If one site has alignment requirements, add compliance at the other

Figure 5.5 Fixing features in opposition

5.1.3 The Constraint Worksheet

Designers without an intuitive or comfortable understanding of the subject need a way of teaching themselves about constraint in the snap-fit interface and understanding its effects on the attachment. A manual approach to documenting constraint using a worksheet as a learning tool can help in this regard. As an interface is developed and evaluated, the designer can use the constraint worksheet to understand the interactions of the interface features and make decisions that lead to optimization of the interface. Use of the worksheet for a short time will help to improve one's understanding of constraint as well as spatial reasoning and the ability to design fundamentally sound and reliable snap-fits [1].

The worksheets shown here are labeled and marked for illustrative purposes. A blank worksheet is provided in Chapter 7. That worksheet can be copied and enlarged for use by the reader. Use the constraint worksheet to evaluate several existing designs before using it during development of a new application. The following steps explain how to use the worksheet. Table 5.1 is labeled so the reader can follow along.

Teaching oneself about constraint can be tedious, largely depending on how intuitive one finds the concept of constraint. If understanding constraint does not come easily, the only way to learn is to struggle with it. Learning in a small group where constraint issues can be debated and discussed is generally more effective than trying to learn it alone. The constraint worksheet and the steps that follow are a starting point for learning. Different readers may prefer another approach and should feel free to modify the process until it is comfortable for

Table 5.1 Worksheet for Tracking Constraint in the Attachment

Interface Requirements		Degrees of Motion											
		Translation						Rotation					
		+x	−x	+y	−y	+z	−z	+x	−x	+y	−y	+z	−z
① Identify desirable axes for mating part stability, translation and rotation								(y axis) (z axis)	(x axis) (y axis)	(x axis) (z axis)		(x axis) (y axis)	
Uni-directional effects	F_M resulting from accelerations and part mass												
	F_F resulting from functional loads												
	② F_N resulting from non-standard loads												
	Engage Direction (ED) & Assembly Force (F_A)												
	Separation Force (F_S)												
Bi-directional effects	Thermal expansion / contraction												
	③ Alignment directions												
	Part compliance												
Constraint pairs ④		⑤	⑥	⑦	⑧	⑨	⑩				⑪		
Resolve bi-directional requirements	Part-to-part alignment												
	Compliance sites	⑦	⑧		⑨		⑩						
	Fine-tuning sites												

X = difficult or not available
A = available
✓ = necessary or required

them. The important result is that the reader understands constraint and can recognize constraint violations in a snap-fit attachment.

1. Recalling the discussion of perfect constraint and part stability, identify desirable directions for planar and linear constraint from the standpoint of part geometry. While it is not always possible to place constraint features at the best or most effective points, it is important to avoid putting them at the least effective points. On the translation side, this involves marking the two more desirable axes for three point constraint and the two more desirable axes for two point constraint. (Circle or highlight the appropriate columns). On the rotation side of the worksheet, the possible directions for rotational constraint should also be identified. These identifications may help when choosing between constraint alternatives later in the process.
   - The distance between constraint pairs (with parallel strength vectors) affects both mechanical advantage against forces and dimensional sensitivity.
   - Remember, as these constraint pairs are moved farther apart, their effectiveness increases.
2. Identify all the force effects that must considered when establishing interface requirements. As a rule these will only be translational effects so the rotation side of the worksheet is not used. The user is free to select any sign convention they choose, but in general, the sign convention should be based on restraint of the mating part. Force effects may include any or all of the following:
   - All forces in the interface due to significant load inputs to the application.
   - Engage direction and assembly force.
   - Separation force.
3. Identify all the bi-directional effects to be considered when establishing interface requirements. Like forces these can generally be expressed as translational effects so the rotation side of the worksheet is not used. These are effects that, by definition, will have consequences in both directions along a given axis. They include:
   - Thermal expansion/contraction.
   - Alignment requirements.
   - Part compliance.
4. List all of the constraint pairs. They can be listed in any order, but the preferred and easiest order to work with is:
   - First list all the locator pairs that establish the interface plane. This is the three-point or planar orientation from the perfect constraint example.
   - Next list the locator pairs that establish linear (two-point) restraint.
   - Next list the locator pairs that establish single point restraint.
   - Finally, list the lock pairs.
   - Label all natural locators with an 'N' as a reminder that they may require special attention if they are to be used as fine-tuning or compliance sites.
5. Identify the contribution each constraint pair makes to removing translational degrees of motion. Work across the top of the worksheet using the six columns of translation. The reader may wish to experiment with two ways to do this and choose the one that works best for them: (1) Constraint pairs are considered one at a time and all DOM removed by

that pair are identified or (2) each DOM is considered and the contribution of each constraint pair (if any) to that DOM is identified.
- Use fractions to indicate the estimated contribution of pairs acting in parallel and having the same sense. It is convenient and generally accurate to assume equivalent strength and stiffness, thus equivalent contribution. If, for example, a panel is held in place by eight lock pairs acting in parallel, each pair would receive a value of $\frac{1}{8}$ in the appropriate cell.
- Check for translational under or over-constraint by studying the entries in the columns. Columns with a total less than one are under-constrained. Columns with a total of one are properly constrained. Columns with a total greater than one may be over-constrained; check the constraint pairs against the rules for proper constraint.
- If an under-constraint condition exists, fix it and adjust the worksheet accordingly.
- If over-constraint due to constraint pair redundancy exists, fix it by removing the least efficient pair (for mechanical advantage and dimensional robustness) and adjust the worksheet accordingly.
- If over-constraint due to constraint pair opposition exists, fix it if possible and adjust the worksheet accordingly, or record the condition for later review. Note the need for feature compliance along that axis.
- One way to fix over-constraint due to opposition is by removing both directions of movement within one constraint pair (most preferred solution). Another is by providing feature compliance at one of the constraint sites.
- Where compliance cannot be used or will not be effective, close tolerances between the opposing constraint pairs will be necessary, but this is the least preferred solution. Evaluate the effects of relative thermal expansion/contraction of the parts and the possibility of warpage or damage to features.
- Identify the primary constraint pair based on the alignment and/or strength requirements of the application. Plan to use this pair as the datum for locating all other constraint features in the interface.

6. Identify all translational directions and the corresponding constraint pairs that:
   - Require high strength to resist interface forces. Mark them with an 'F'.
   - Require positional accuracy to satisfy alignment requirements. Mark them with an 'A'.
   - If strength or alignment requirements are identified in both directions along the same axis, over-constraint in opposition should be avoided along that axis because it cannot be fixed using compliance. If over-constraint in opposition was noted in Step 5, it must be fixed.

7. Identify all translational directions and corresponding constraint pairs where feature compliance can be added. Mark them with a 'C'.
   - While these may be at the same constraint pair, they must not be in a column marked with an 'F' or an 'A'. Compliance sites should not carry forces or provide critical alignment.

8. Identify translational directions and corresponding constraint pairs where expansion and contraction due to thermal effects may occur. Mark them with a 'T'.
   - In these directions, over-constraint in opposition should be avoided if possible. Otherwise, compliance in one of those directions will be required. Mark with a 'C'.

9. Identify sites where fine-tuning enablers may be used. Mark them with an 'E'. These must include sites and directions marked with an 'F' or an 'A'.

- There should be fine-tuning sites in each of the three translational directions, but not in opposing directions. For example, combinations like (+x, +y, −z) or (−x, +y, −z) are OK. A combination like (+x, −x, +y, +z) is not OK.
- Fine-tuning sites should control all critical alignment directions.

10. Identify directions in which *part* compliance is an issue. Note that this is not the same as *feature* compliance.

- Highly compliant parts (soft or flexible parts like panels) may require multiple constraint pairs (acting in parallel) to remove all possible flexure. Part compliance is often an issue in parts with the *panel* basic shape.
- Verify these constraint pairs are properly spaced to ensure against part flexure.
- Adding stiffening features such as ribs to increase part stiffness is often desirable.

11. Identify the contribution each constraint pair makes to removing rotational degrees of motion. Work across the top of the worksheet using the six columns of rotation.

- Rotation is removed through constraint pairs acting as couples. A single constraint pair of sufficient length can also act as a couple; a very long wedge to slot is an example.
- Use fractions to indicate the estimated contribution of each pair. Assume equivalent strength and stiffness.
- Note that, similar to the effect described in #4 above, each couple involves strength vectors in parallel, but as a couple, they will be acting in opposite directions.
- As with constraint pairs acting in parallel to prevent translational movement, effectiveness in strength and dimensional stability increases as the distance between the constraint pairs increases.
- Check to verify there is no over or under-constraint in rotation. If there is, fix it and adjust the worksheet. Verify you have not changed any translational constraint conditions.

As the reader will quickly discover in trying to actually evaluate constraint and the feature interactions, it is very much an iterative procedure. Do not expect to experience a rigorous thought process that will lead to a final answer in just one pass through the evaluation process.

Table 5.2 shows how the worksheet could be filled out for the perfect constraint example introduced earlier in the chapter with an external force and a location requirement added as shown in Fig. 5.6. Table 5.3 shows how the worksheet could be filled out for the simple but more realistic application in Fig. 5.7. This application is a slight variation of the switch application shown in Fig. 4.4. The Chapter 4 application is over-constrained in rotation around the z-axis. The reader might want to evaluate that application using the worksheet to see how the rotational over-constraint is exposed. Again, have some parts (i.e. "models") in hand to help visualize part behaviors if you intend to work through these examples using the worksheet.

Table 5.2 Example of Using The Worksheet—Perfectly Constrained Object Shown in Fig. 5.6

Interface Requirements			Degrees of Motion										
			Translation						Rotation				
			+x	−x	+y	−y	+z	−z	+x	−x	+y	−y	+z	−z
Uni-directional effects	Identify desirable axes for mating part stability, translation and rotation				Linear		Planar		(y axis) (z axis)	(x axis)	(x axis) (z axis)	(x axis)	(x axis) (y axis)	
	F_M resulting from accelerations and part mass													
	F_F resulting from functional loads				(F_F)									
	F_N resulting non-standard loads						Not applicable							
	Engage Direction (ED) & Assembly Force (F_A)													
	Separation Force (F_S)													
Bi-directional effects	Thermal expansion / contraction		(No gap)											
	Alignment directions			(1)		1								
	Part compliance					1								
Constraint pairs	(a) surface (N) to surface (N)		1	1	1	1	1	1	1/2	1/2	1/2	1/2	1/2	1/2
	(b) edge (N) to wall													
	(c) edge (N) to catch													
	(d) locator to edge (N) (must provide alignment)													
	(e) locator to edge (N) (must carry force)													
	(f) lock to surface (N) (must provide compliance)													
	totals		1	1	1	1	1	1	1/2	1	1/2	1	1/2	1
Resolve bi-directional requirements	X = difficult or not available A = available** ✓ = necessary or required	Part-to-part alignment	✓	X	✓	A	A	X*	1	1	1	1	1	1
		Compliance sites	✓	✓	A	A	A	X*						
		Fine-tuning sites	X	✓	✓	A	A	X*						

* Difficult where both features are natural locators.

Select sites for compliance and fine-tuning along the y and z axes from the available choices.

5.1 The Importance of Constraint   149

Table 5.3 Example of Using The Worksheet—Simple Part Application Shown in Fig. 5.7

Interface Requirements		Degrees of Motion											
		Translation						Rotation					
		+x	−x	+y	−y	+z	−z	+x	−x	+y	−y	+z	−z
Identify desirable axes for mating part stability, translation and rotation				Linear		Planar		(y axis) (z axis)	(y axis) (z axis)	(x axis) (z axis)		(x axis) (y axis)	(x axis) (y axis)
Uni-directional effects	F_M resulting from accelerations and part mass												
	F_F resulting from functional loads						F_F						
	F_N resulting non-standard loads						F_A						
	Engage Direction (ED) & Assembly Force (F_A)												
	Separation Force (F_S)					F_S							
Bi-directional effects	Thermal expansion / contraction												
	Alignment directions			Consistent gap		Flush							
	Part compliance												
Constraint pairs	(a) surface (N) to surface (N) (both are natural)						1						
	(b) land to edge (N)			1/2				1/2	1/2	1/2	1/2	1/2	1/2
	(c) land to edge (N)			1/2								1/2	1/2
	(d) land to edge (N)				1/2								
	(e) land to edge (N)				1/2								
	(f) land to edge (N)		1										
	(g) land to edge (N)	1											
	(h) lock to edge (N)					1/2		1/2	1/2	1/4	1/4		
	(i) lock to edge (N)					1/2				1/4	1/4		
Resolve bi-directional requirements	Part-to-part alignment	✓	✓	✓	✓	✓	✓						
	Compliance sites*	X	X	X	X	X	X						
	Fine-tuning sites**	✓	✓	✓	✓	✓	✓						

X = difficult or not available
A = available
✓ = necessary or required

* Because of alignment requirements, compliance is difficult at any site.
** If determined necessary, one of the surface-to-surface natural locators could be changed to lands to support fine-tuning.

150   Other Snap-Fit Concepts   [Refs. on p. 161]

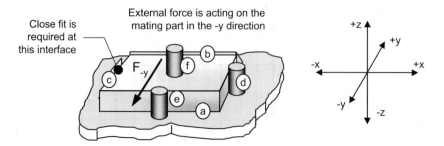

Figure 5.6 A "perfectly" constrained snap-fit with some requirements illustrated for the constraint worksheet example in Table 5.2

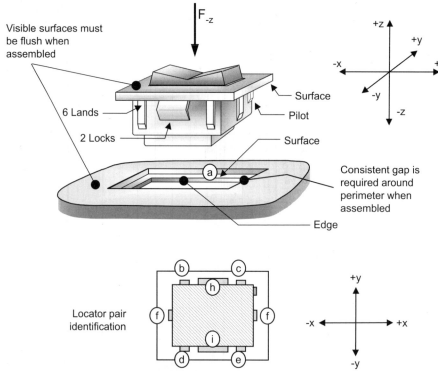

Figure 5.7 A solid to opening application illustrated for the constraint worksheet example in Table 5.3

5.1.4 Additional Comments on Constraint

Remember that a blank copy of the constraint worksheet is included in Chapter 7.

A field that makes extensive use of constraint principles is part fixturing. When fixtures are developed to hold and locate parts for machining or for dimensional checking, proper constraint is essential.

Most of the constraint principles expressed in this section as well as in Chapters 2 and 3 in the form of qualitative design rules also lend themselves to expression in mathematical terms. Tools for optimizing a snap-fit interface in terms of constraint, strength, compliance and tolerances can and should be developed [2].

5.2 Lock Decoupling

Lock decoupling is the degree to which a lock's retention behavior is independent of its assembly behavior. Understanding decoupling and the additional lock design options it provides will help the designer improve lock designs and resolve lock performance problems. This section describes lock feature decoupling in detail using the common cantilever hook as an example.

5.2.1 The Lock Feature Paradox

By their nature, locking features present a design paradox. They should be easy to assemble, i.e. *weak*. (Unless automatic or robotic assembly is planned, in which case assembly force levels may be less important.) But locks must also be strong to resist breakage or unintended separation. This conflict between weak and strong performance requirements can sometimes force design compromises that don't adequately satisfy either requirement. A solution to this dilemma is made possible by decoupling (separating) the lock's assembly performance from its retention/separation performance.

5.2.2 Decoupling Examples

The concept of decoupling isn't difficult to understand and it can be a powerful tool for solving design problems. Decoupling is best introduced with a few examples.

Imagine buying a ladder that is the correct length for climbing to the roof of a house. Later you decide to wash the windows of the house, but the ladder is too long for that job. You must move the ladder's base far away from the house to get the ladder's end down to window level. With the base so far away, you can't climb the ladder without it sliding down the wall. The ladder's useful height (H) is limited by the safe distance (D) of the ladder's base from the house. D and H are strongly *coupled* because we cannot change H without affecting D, Fig. 5.8a.

Obviously you made a mistake buying *that* ladder; you should have looked for one with working height and safe distance decoupled. "HA!" you're thinking, "Who has time for fancy engineering concepts when we're fixing a roof and washing windows? Besides I can just imagine the look I would get if I told a hardware store clerk that I wanted some kind of exotic *decoupled* ladder." In reality, decoupling is an everyday occurrence. The common extension ladder is designed to decouple D and H, Fig. 5.8b. Most people who buy a ladder instinctively consider decoupling without even realizing it.

Another decoupling example is a door and doorknob. The door closes and latches with a simple push. But simply pulling on the door or on the doorknob will not open the door. The

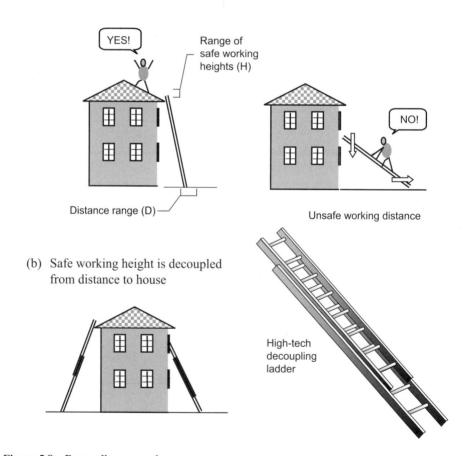

Figure 5.8   Decoupling example

doorknob must be *turned* to release the latch. The door's latching and retention characteristics are decoupled: a push assembly motion and a rotational release motion.

5.2.3 Levels of Decoupling

There are a number of ways that snap-fit locks can be decoupled and they can be broken down into four "levels". The levels are defined according to how the decoupled assembly and retention behaviors are analyzed. The levels can also be ranked by effectiveness. This is important because the more effective the decoupling, the stronger we can make the lock in retention relative to the assembly force.

Because the cantilever hook is most sensitive to the linkage between assembly and retention, we will use it to explain decoupling. In a typical cantilever hook (beam and catch) we find that assembly behavior is affected by beam bending, friction, catch height and the insertion face angle. In the same hook, retention behavior is affected by beam bending, friction, catch height and the retention face angle. Thus, both assembly and retention are directly related to beam bending, friction and catch height. The insertion face angle, however, only affects assembly and the retention face angle only affects retention.

In the common cantilever hook, any changes made to the beam will affect both assembly and retention. Making the beam thicker for more strength also increases the strain at its base during assembly deflection and increases the required force for assembly, making it harder to assemble. If, on the other hand, the beam is made thin for easy assembly, it becomes weak. We can see these relations in the basic calculations for assembly force (F_A) and retention force (F_R).

5.2.3.1 No Decoupling (Level 0)

There is no decoupling in the hook shown in Fig. 5.9. Assembly and retention behavior are virtually identical because:

- The calculations (bending) used to analyze assembly and retention are identical because the hook behavior is the same: beam bending.

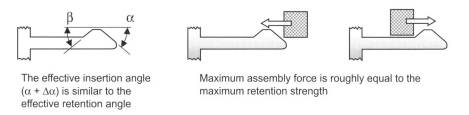

The effective insertion angle ($\alpha + \Delta\alpha$) is similar to the effective retention angle

Maximum assembly force is roughly equal to the maximum retention strength

Figure 5.9 A cantilever hook with no decoupling (Level 0)

154   Other Snap-Fit Concepts   [Refs. on p. 161]

- The same variable (angle) is used in the calculations, the only difference being that the insertion face angle (α) is used in the assembly calculation and the retention face angle (β) is used in the retention strength calculation.
- The variables (α and β) have the same values: ($\alpha = \beta$).

(For sake of discussion, we will ignore the relatively small change in the moment arm of the deflection force as it moves over the insertion and retention faces of the catch. We will also ignore the change in the face angles as the beam deflects.)

5.2.3.2   Level 1 Decoupling

For the hook shown in Fig. 5.10, the only independent variables we have to work with to adjust assembly and retention behavior are the insertion and retention faces. By making the insertion face angle (α) low, we can get *lower* assembly forces but no matter what we do to the insertion angle, assembly force reduction will ultimately be limited by the bending strength of the beam. The same is true for retention behavior. We can improve retention by increasing the retention face angle (β) but again, we are limited by the beam's bending behavior.

We could also try to do something with friction on the insertion and retention faces but we are still limited by the beam's behavior. Thus, for the common cantilever hook, assembly and retention behavior ultimately depend on the bending behavior of the beam. The lock in Fig. 5.10 is *partially* (but rather ineffectively) decoupled at the insertion and retention faces. In conclusion, we can point out that this style of cantilever hook is a relatively poor performer with respect to assembly and retention decoupling and it cannot be made any

(a) Adjusting the retention face angle to increase retention force

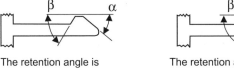

The retention angle is greater than the insertion angle

The retention angle is much greater than the insertion angle

(b) Adjusting the insertion face to face angle to decrease insertion force

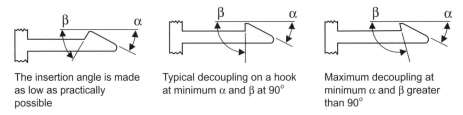

The insertion angle is made as low as practically possible

Typical decoupling on a hook at minimum α and β at 90°

Maximum decoupling at minimum α and β greater than 90°

Figure 5.10   Variations of Level 1 decoupling in a cantilever hook

better. This is Level 1 decoupling and it is the lowest and least effective level of decoupling. It is also the easiest and the most common. The degree of Level 1 decoupling will determine whether the lock is releasing or non-releasing with the retention face angle being the determining factor.

5.2.3.3 Level 2 Decoupling

A significant increase in decoupling effectiveness occurs when we move to Level 2 decoupling as illustrated by the side-action hooks in Fig. 5.11. Simply turning the retention mechanism 90° causes major changes in the behavior of the hook.

The equation for bending force (F_P) in a cantilever beam is:

$$F_P = \frac{wt^2 E \varepsilon}{6L} \quad (5.1)$$

Where: w is beam width; t is beam thickness; ε is strain; L is beam length; E is the material's modulus of elasticity.

In this equation, we see that F_P is directly proportional to beam width (w) and to beam thickness (t) squared. For the beam shown in Fig. 5.12a, the cross-section is shown with measurements $t = 1$ and $w = 5$. For determining assembly behavior, we would use these values in the calculation for assembly force and the value of the expression (wt^2) is ($5 \times 1^2 = 5$).

For retention strength, however, the variables change, Fig. 5.12b. The dimension that was previously beam width is now beam thickness ($t = 5$) and the dimension that was thickness is now width ($w = 1$). For these new values the value of the expression (wt^2) is ($1 \times 5^2 = 25$). By this calculation, the retention strength of this side action hook could be as much as 5 times the strength of a similar hook with Level 1 decoupling. (Beam distortion is possible and the actual effect may not be as high as 5×, but the improvement is still significant.) The only change has been to turn the catch sideways.

Level 2 decoupling occurs when different *variables* are used in the equations. In the side-action hook, turning the catch 90° on the beam causes the beam width (w) and thickness (t) variables to change in the equations. However, the same (bending) equations are used, and they are applied to the same feature (the beam). Level 2 locks may be releasing or non-releasing.

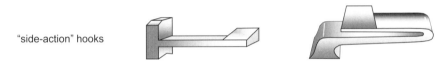

"side-action" hooks

Figure 5.11 Level 2 decoupling. Turning the retention feature on a cantilever hook 90° can make a significant difference in performance

(a) The beam bends around the thinner section during assembly

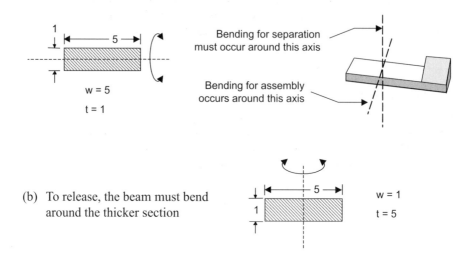

(b) To release, the beam must bend around the thicker section

Figure 5.12 Effect of Level 2 decoupling on hook performance

5.2.3.4 Level 3 Decoupling

Level 3 decoupling occurs when different assembly and retention behaviors within the same feature require different equations for evaluation. This gives even greater independence between the assembly and retention behaviors, thus increasing the designer's control over each of them. Level 3 decoupling occurs naturally in the trap lock, Fig. 5.13a, where assembly involves beam bending and is evaluated using the equations of bending. Retention however must be evaluated with equations for beam behavior under axial compression. Recall that trap locks are one of the more desirable lock features for ease of assembly and retention strength.

Another example of Level 3 decoupling is shown in Fig. 5.13b. In this loop-catch lock pair, insertion involves beam bending, but retention involves material shear and tension. Again, retention is evaluated with different calculations than assembly. With the loop, the level of decoupling (1 or 3) depends on the angle of the retention face on the catch.

In general, locks having Level 3 decoupling are inherently stronger than those with Level 1 or 2 decoupling.

5.2.3.5 Level 4 Decoupling

Level 4 decoupling involves the use of different features for assembly and retention and dramatic differences in assembly and retention performance are possible. In the example, thin and flexible locks on the mating part engage through a hole in a surface of the base part, Fig. 5.14a. (Maybe the mating part material is relatively rigid and will not tolerate high strains.) Once the mating part is in place, a pin is pushed into the mating part. The pin-to-

(a) Traps, by definition, are Level 3 decoupling

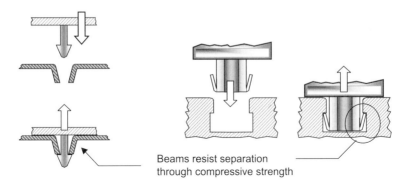

Beams resist separation through compressive strength

(b) A loop is Level 3 if resistance to release is tension and shear

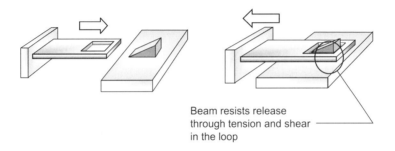

Beam resists release through tension and shear in the loop

Figure 5.13 Level 3 decoupling

mating part attachment is normally a snap-fit or a press-fit. When installed, the pin prevents the hooks from deflecting and releasing. Retention strength of the mating part to base part attachment can be very high and is a function of the tensile strength of the mating part material and the cross-sectional area of the hook beams. Figure 5.14b shows how, in another application, a feature on a third part can take the place of the pin. Level 4 locks are non-releasing.

Another example of Level 4 decoupling is shown in a solid-surface application, Fig. 5.15. In this application, a solid is located to a surface using lugs arranged around its perimeter. The lugs are inserted into the holes in the surface and the solid is slid against the surface so that each lug moves into the narrow area of the hole. No locking occurs as the lugs engage. Once the solid is in position, the bezel is placed over the solid and pushed into engagement with the surface. The strong pins on the bezel fill the holes behind the lugs to prevent the solid from sliding and the hooks on the bezel hold it in place to the surface. Retention strength for the solid to the surface can be very high because it is a function of the strength of the pins and the lugs.

(a) A separate pin fills the space between the locks to prevent deflection and release

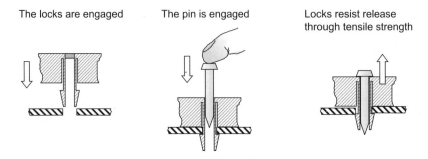

(b) A feature on a third part prevents the lock from deflecting

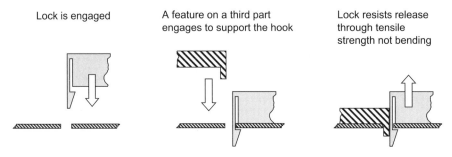

Figure 5.14 Level 4 decoupling

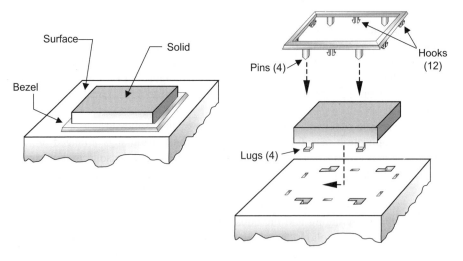

Figure 5.15 Level 4 decoupling in a bezel application

The author's observation has been that any application employing a bezel as a closure around or over a gap between the mating and base parts is a good candidate for Level 4 decoupling. In some applications, addition of a bezel to allow implementation of Level 4 decoupling is probably a cost-effective solution to an attachment situation.

Level 4 decoupling involves use of different features for assembly and retention and is the highest form of decoupling. Dramatic differences in assembly and retention performance are possible. The *push-pin* style plastic fastener (discussed in Chapter 7 as a substitute when an integral lock feature will not work) employs Level 4 decoupling.

5.2.4 Decoupling Summary

Recall that one of the performance enhancements discussed in Chapter 4 was "retainers". Do not confuse decoupling effects with retainers. Although a retainer enhancement will improve the retention strength of an attachment, it also increases the assembly force so the effects are not independent, Fig. 5.16. By our definition, the retainer is not decoupling.

Table 5.4 summarizes the distinctions between the five levels of decoupling. Also, recall the discussion of lock efficiency in Chapter 3. Lock efficiency is the ratio of a lock's retention strength to its assembly force. Retainer enhancements can improve lock efficiency, but the higher levels of decoupling are by far the most useful and effective way to improve lock efficiency.

- Level 0 (no decoupling) was illustrated by a hook with *equivalent* retention and insertion face angles.

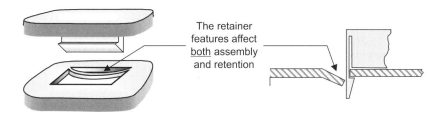

Figure 5.16 Retainer enhancements are not decoupling

Table 5.4 Decoupling Summary

Potential lock efficiency	Level	Features	Equations	Variables	Values
Lowest	0 (none)	same	same	same	same
	1	same	same	same	*different*
	2	same	same	*different*	N/A
	3	same	*different*	N/A	N/A
Highest	4	*different*	*different*	N/A	N/A

- Level 1 decoupling occurred when the hook's insertion face angle was decreased for lower assembly force and the retention face angle increased for higher separation force. Only the *values* of the face angle variables changed.
- Level 2 decoupling occurred when different *variables* were used in the equations. In the side-action hook, turning the catch 90° on the beam causes the beam width (w) and thickness (t) variables to change in the equations.
- Level 3 decoupling occurred when completely different assembly and retention behaviors required different *calculations*.
- Level 4 decoupling required the use of different *features* for assembly and retention.

5.3 Summary

This chapter provided additional discussion of two important concepts in snap-fit design. Constraint has already been discussed in other chapters, but additional issues were discussed and a worksheet was introduced for tracking how degrees-of-motion are removed by constraint pairs.

Decoupling as a way of improving lock performance was introduced and discussed in detail. Decoupling provides additional design options when balancing lock performance trade-offs between assembly and retention behavior.

5.3.1 Important Points in Chapter 5

- Do not rely on clamp load in a snap-fit and do not try to design clamp load into a system of plastic parts. Design instead for a line-to-line fit.
- Conscious or explicit consideration of constraint in attachments is not common practice because many designers are accustomed to specifying threaded fasteners.
- If you do not have a high comfort level with your understanding of constraint, use the constraint worksheet until you do.
- Perfect constraint is a theoretical ideal. By avoiding constraint mistakes and minimizing some non-preferred conditions, the designer can ensure a snap-fit with proper constraint. *Proper* constraint is essentially the absence of *improper* constraint.
- Lock decoupling is the degree to which a lock's *retention* behavior is independent of its *assembly* behavior.
- Although both can improve a lock feature's efficiency, retainer enhancements and decoupling are not the same thing.

5.3.2 Design Rules Introduced in Chapter 5

- Over-constraint due to opposing constraint pairs is not recommended, but is often a practical necessity. To compensate, use compliance enhancements.

- Over-constraint due to redundant constraint pairs is inefficient.
- An under-constraint condition is unacceptable and must be fixed.
- The common cantilever hook lock is limited in its decoupling ability and caution in its use is recommended.
- The loop style cantilever lock, the side-action hook and the trap are all higher level locking devices than the hook and are preferred.
- Applications involving bezels lend themselves to Level 4 decoupling.

References

1. Luscher, A.F., Bonenberger, P.R., 1997, "Part Nesting as a Plastic Snap-fit Attachment Strategy", DETC97/DTM-3893, *Proceedings of DETC '97, ASME Design Engineering Technical Conference*, September 1997.
2. Bonenberger, P.R., 1995a, "A New Design Methodology for Integral Attachments", ANTEC '95 Conference of the Society of Plastics Engineers, May 1995, Boston, MA.

6 Feature Design and Analysis

Final design and analysis of the constraint features is appropriate only after a fundamentally sound attachment concept has been created. Locator features, in most cases, require little analytical attention because they are strong and inflexible. Unlike locks, locators are not required to balance assembly deflection and strain with retention strength. If evaluation of locator strength is required, it normally involves straightforward and simple shear or compression strength calculations. For that reason, locator feature calculations are not discussed in this chapter.

Generally, issues of feature design and analysis focus on locking features because, by their nature, locks are more complex and have more performance requirements. Lock feature calculations also represent the traditional "feature level" of snap-fit technical understanding. As such, there is a long history of writing on the subject. Many publications related to lock calculations are listed at the end of this chapter. Specific publications are referenced in the text where they contain unique or particularly useful information.

Because of the volume and availability of feature level information, this chapter does not go deeply into the subject, nor does it go too deep into the materials issues associated with feature analysis. Many excellent sources of information exist on that subject; some of them are also listed at the end of this chapter.

This chapter will briefly discuss some data and materials issues and present the common calculations for cantilever hook behavior. Adjustments for greater accuracy of these calculations are also discussed. Some calculations for other styles of lock are also introduced.

In some instances, feature calculations are not deemed necessary or practical for initial part development. The designer simply wishes to produce a lock feature that has a good chance of working and is reasonably representative of a final feature design. To support this approach, some rules of thumb for initial cantilever lock design are provided.

In preparation for feature analysis, designers must identify the specific purpose of the analysis. They must also have access to the data necessary to characterize the material properties of the part in question. Design and analysis normally involves evaluating snap-fit constraint features for any or all of the following:

- Assembly force
- Assembly strain
- Retention strength
- Release forces
- Release strains

For some of the above, analysis requires property data for fresh materials. For others, data for fresh as well as for aged material may be needed.

6.1 Pre-Conditions for Feature Analysis

For an accurate and valid feature analysis, certain pre-conditions must be met. Feature analysis should occur only after proper interface constraint is verified. Proper constraint ensures that forces in the attachment are statically determinate and that only the intended forces (forces to be considered during analysis) act on the snap-fit features. Constraint was introduced in Chapter 2 and discussed in more detail in Chapters 3 and 5.

The interface design should be as dimensionally robust as possible. Primary and secondary datum sites on both parts should be selected with the constraint features in mind. Ideally, the most dimensionally critical locator pair will be the datum for all other constraint features. This reduces the need for close tolerances and means that forces on the features are more predictable. Always keep in mind the three requirements that apply to analysis: strength, constraint and robustness. Of course, the ultimate goal of analysis is ensuring feature strength.

6.2 Material Property Data Needed for Analysis

The intention here is not to make the reader an expert in polymers. The purpose is to introduce some important concepts for basic understanding and encourage the reader to ask intelligent questions of the resin supplier and/or their polymers expert. Many times, a material is selected for a particular application based on appearance and functional requirements. Snap-fit feature performance is not a prime consideration. The snap-fit must be made to work with the given material. An appreciation of some materials issues will help the designer recognize potentially difficult situations early in the development process and know some of the questions to ask of a polymer expert.

Four material properties normally appear in feature analysis calculations. They are: stress (σ) or strain (ε), modulus of elasticity (E) and coefficient of friction (μ). The earlier in the development process that the designer has information about these properties, the better. For most snap-fit performance calculations, strain data is preferred over stress.

6.2.1 Sources of Materials Data

Stress-strain and related information may be found in several forms and some are more useful than others. Material information in *product brochures* is appropriate only for general product comparison or for initial screening of products for an application. It should not be used for part design or for snap-fit feature analysis.

Material data sheets also represent the supplier's interpretation of laboratory data. Data sheets are more detailed and useful than brochure information but are, of necessity, based on general use assumptions and specific test conditions. They can only provide data at specific points (single point data) and their creation is subject to normal differences of data interpretation. If used for analysis, ensure the data represents the information needed (a supplier's terminology may not be the same as yours) and that you fully understand the conditions under which the data was generated. Be aware that test and sample preparation procedures may differ between suppliers.

Materials encyclopedias, supplier databases and universal databases contain information similar to that in the materials data sheets. However, different test and sample preparation methods may make direct comparisons difficult. An exception is the CAMPUS® database [1]. CAMPUS® stands for Computer Aided Material Preselection by Uniform Standards. The CAMPUS® database contains data from over 45 plastics producers, including stress-strain curves and other mechanical, thermal and electrical properties. One of its primary attractions is that the data is based on uniform testing to ISO Standards, making direct comparisons of material properties possible. The database is available to qualified customers of the member companies.

Stress-strain curves are the preferred form of data for snap-fit design and analysis. The stress-strain curves allow the designer and the materials expert to interpret the data as they see fit for a particular application. The designer must still verify that the conditions under which the data was generated represent the application. Sometimes, the designer may need to request that these curves be generated for a particular set of conditions.

Much of the published stress-strain information is based on tensile testing. Tensile test data is desirable for tensile loading conditions and acceptable for other conditions when no other is available, but data generated in tests that represent actual loading conditions is preferred. For some snap-fit feature analysis, bending is the primary mode of deflection. For this situation, stress-strain curves generated by flexural testing would be preferred. For shear conditions, data generated through shear testing is desirable.

Although stress-strain curves created to closely match the intended use of the material in the application are preferred, they should also be used with caution. The following quote is from one resin supplier's design guide [2] but it is applicable to data from all suppliers.

> Values shown are based on testing of laboratory test specimens and represent data that fall within the standard range of properties for natural material.... These values are not intended for use in establishing maximum, minimum or ranges of values for specification purposes. [The user] must assure themselves that the material as subsequently processed meets the needs of their particular product or use.

This means no matter how good the material property data is, it is the result of laboratory testing under standard conditions. These conditions cannot represent all the variables and conditions of a particular application. End use testing of production parts is required to verify performance.

6.2.2 Assumptions for Analysis

Analysis calculations for plastic, unless otherwise noted, are based on three assumptions about the material. These are: elastic linearity, homogeneity and isotropy. In reality, plastics do not meet these assumptions although some plastics come closer than others. These assumptions are necessary however, if we are to apply relatively simple calculations using traditional equations of structural analysis and they are reasonable for most snap-fit analysis. One reason these assumptions are acceptable is that, in most cases, predictive analysis of snap-fit behavior is not an exact science. The effect of these assumptions on the accuracy of the analysis is not as significant as the effects of many other variables on the calculations. Some of these other effects are discussed in this chapter.

The plastic is *linearly elastic*. The stress-strain curve is linear in the area of analysis. (The opposite of elasticity is plasticity.) In reality, most plastics are not linear over the useful area of their stress-strain curve. We compensate for this by assuming a linear stress-strain relationship (the secant modulus) for the range of stress and strain in which we are working.

The plastic is homogeneous. The material's composition is consistent throughout the part and a small piece of the part will have the same physical properties as the whole part. (The opposite of homogeneity is heterogeneity.) In reality, plastic part composition depends on many factors, including raw material mixing, mold flow and cooling. Proper mold and part feature design can help ensure that material properties in the areas of analysis are reasonably close to the predicted properties. Safety factors and conservative calculations can also compensate.

The plastic is isotropic. The physical properties at any point in the material are the same regardless of the direction in which the sample is tested. (The opposite of isotropy is anisotropy.) In reality, filled and glass reinforced materials in particular do not exhibit isotropic behavior. Sometimes data for these materials will indicate the direction of testing. Sometimes the data will only reflect the maximum performance direction. Proper part and mold design helps ensure that the high performance properties are oriented in the correct directions in the final part. Values used in analysis should reflect anisotropic behavior if it exists.

6.2.3 The Stress-Strain Curve

The most important information needed for analysis is a material's stress-strain relationship. The best way to show this relationship is in a stress-strain curve, a graph of stress vs. strain for a material under a given set of laboratory test conditions, Fig. 6.1. The initial modulus is the slope of the stress-strain curve at relatively low stresses and strains. It is a tangent to the initial portion of the curve. If the plastic exhibits some linear behavior, the initial modulus will be the slope up to the proportional limit. Or, if the initial portion is non-linear, the initial modulus may be reported as a secant modulus, usually at 1% strain.

Unlike the characteristic stress-strain curve for steel, the shape of which is representative of all steels, stress-strain curves for plastics may be quite different from material to material. They are also subject to interpretation. This is why it is good to get stress-strain curves for materials when doing an analysis. Some typical stress-strain curves for plastics are shown in Fig. 6.2 and the important points on each curve are defined. Not every point will show up on every curve. Figure 6.2 and the definitions of terms that follow are from *Designing with Plastic—The Fundamentals, Design Manual TDM-1*, courtesy of Ticona LLC [2]. In the

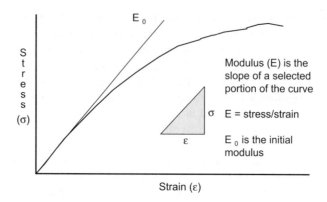

Figure 6.1 The basic stress-strain curve

author's opinion, this document is an excellent blend of material and design information for the snap-fit designer who is not a polymer expert. It is highly recommended.

Proportional limit (A). With most materials, some point exists on the stress-strain curve where the slope begins to change and the linearity ends. The proportional limit is the greatest stress at which a material is capable of sustaining the applied load without deviating from the proportionality of stress to strain. This limit is expressed as a pressure in MPa (or in psi) and is shown as Point *A* in Fig. 6.2. Note that some materials maintain this proportionality for large measures of stress and strain while others show little or no proportionality, as previously discussed.

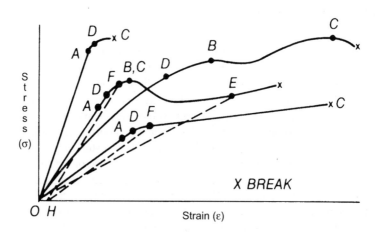

Figure 6.2 Typical stress-strain curve, Courtesy of Ticona LLC, *Designing With Plastic—the Fundamentals*

6.2 Material Property Data Needed for Analysis

Yield point (B). Yield point is the first point on the stress-strain curve where an increase in strain occurs without an increase in stress. The slope of the curve is zero at this point. Note that some materials may not have a yield point.

Ultimate strength (C). The ultimate strength is the maximum stress a material withstands when subjected to an applied load. This is also a pressure expressed in MPa (or psi) and is denoted by Point C.

Elastic limit (D). Many materials may be loaded beyond their proportional limit and still return to zero strain when the load is removed. Other materials, particularly some plastics, have no proportional limit in that no region exists where the stress is proportional to strain (the material obeys Hooke's law). However, these materials may also sustain significant loads and still return to zero strain when the load is removed. In either case, point D on the stress-strain curve, represents the point beyond which the material is permanently deformed if the load is removed. This point is called the elastic limit.

Secant modulus (E). The secant modulus is the ratio of stress to corresponding strain at any point on the stress-strain curve. For instance, in Fig. 6.2 the secant modulus at Point E is the slope of the line OE.

Yield strength (F). Some materials do not exhibit a yield point. For such materials, it is desirable to establish a yield strength by picking a stress level beyond the elastic limit. Although developed for materials that do not exhibit a yield point, this value is often used for plastics that have a very high strain at the yield point to provide a more realistic yield strength. This is shown as Point F on the curves. The yield strength is generally established by constructing a line parallel to OA at a specified offset strain, Point H. The stress where the line intersects the stress-strain curve at point F is the yield strength at H offset. For instance, if Point H were at 2% strain, then Point F would be termed the "yield strength at a 2% strain offset."

The three basic types of plastic stress-strain behavior are shown in Fig. 6.3. Toughness is a measure of a material's resistance to impact loads and is represented by the area under the

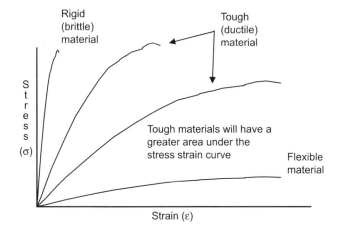

Figure 6.3 Plastic toughness vs. brittleness vs. flexibility

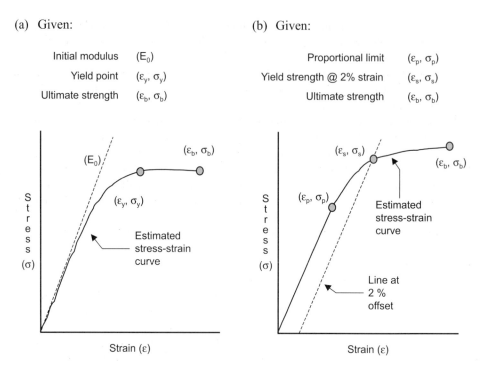

Figure 6.4 Estimating a stress-strain curve from available data

stress-strain curve. Thus, the rigid (brittle) and flexible materials represented here will have lower toughness than the ductile material. Tough plastics are the preferred materials for snap-fits. Snap-fits in brittle plastics require very careful design and analysis with particular caution if impact loads are present in the application. Flexible materials normally do not lend themselves to snap-fits.

The stress-strain curve is an important source of information for feature analysis. If one is not available, a reasonable representation can sometimes be constructed [3] from the information provided on the material data sheet, Fig. 6.4. From this constructed curve, other values needed for analysis can be estimated. Of course, the accuracy of this curve (or any stress-strain curve) must be taken into account when interpreting analysis results. If a stress-strain curve must be estimated, referring to Fig. 6.2 will give the reader a general idea of how various curves might be shaped depending on which data points are available.

6.2.4 Establishing a Design Point

For setting very early or preliminary design targets, the strength values on standard product data sheets can be multiplied by the percentages shown in Table 6.1 [2,3]. Reference [3] also

6.2 Material Property Data Needed for Analysis

Table 6.1 Maximum Strength Estimates for Preliminary Part Design [2, 3]

	When feature failure is not critical	When feature failure is considered critical
For intermittent loading (not cyclic or fatigue loads)	25–50%	10–25%
For constant loads	10–25%	5–10%

Multiply the strength values in material data sheets by these factors for preliminary analysis and product screening. The resulting estimates are not a substitute for final analysis and end-use testing.

provides a much deeper discussion of safety factors and the introduction of other factors reflecting materials and processing effects into determining the design point.

For final analysis, establishing the design point from a stress-strain curve is recommended. The design point represents the maximum stress and strain allowed in the feature being analyzed. The design point also establishes the secant modulus. It may be necessary to determine several design points using several stress-strain curves, each one representing a different condition under which the snap-fit is expected to perform. It may also be necessary to ask the supplier to generate curves representing specific conditions for the application. Conditions for which a design point should be established include both short and long-term considerations.

Typical short-term conditions may be a new/fresh material at room temperature. This would generally be appropriate for evaluating initial assembly behavior unless initial assembly involves temperature extremes or aged material. Typical long-term conditions would comprehend the applied load history, expected number of assembly/disassembly cycles, thermal, ultra-violet and chemical aging effects, material creep properties and ambient temperature effects.

Once stress-strain curves have been obtained, the following guidelines can be used to establish an initial design point for each curve.

a. For Applications Where Strain is Fixed

These are applications in which a feature is deflected during assembly then remains at some level of deflection for the life of the product. This is a long-term loading condition.

- For ductile and high-elongation plastics set the maximum permissible strain at 20% of the yield point or yield strain, whichever is lower, Fig. 6.5a.
- For brittle and low-elongation plastics that don't exhibit yield set the maximum permissible strain at 20% of the strain at break, Fig. 6.5b.

b. For Applications Where Strain is Variable

The assembly process itself involves a change in strain. When deflections occur very rapidly, as in assembly or impact loading, feature analysis should be based on *dynamic strain*, not on stress or static strain. Because of the time-dependence of plastic behavior, it is very possible for calculated stress to exceed the stress at yield without causing damage when deflection

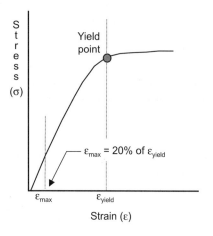 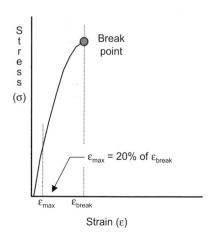

Figure 6.5 Design points for fixed strain applications

occurs rapidly. When loads or deflections occur rapidly, as they do during assembly, use the dynamic strain limit when determining the design point. When loads or deflections occur more slowly, as during disassembly or sustained loading, maximum allowable stress or static strain values can be used in the calculations.

Some suppliers may recommend a maximum working stress level in their material design information. Use stress for evaluating long-term loading conditions.

Two material situations apply in applications with variable strain:

c. Materials With a Definite Yield Point

For a low number of assembly/disassembly cycles (~1–10 cycles), set the maximum permissible strain at 70% of the strain at yield, Fig. 6.6a [3].

For higher assembly/disassembly cycles (>10 cycles), set the maximum permissible strain at 40% of the strain at yield, Fig. 6.6b.

d. Materials Without a Definite Yield Point

For a low number of assembly/disassembly cycles (~1–10 cycles), set the maximum permissible strain at 50% of the strain at break, Fig. 6.7a [3].

For higher assembly/disassembly cycles (>10 cycles), set the maximum permissible strain at 30% of the strain at break, Fig. 6.7b.

e. The Secant Modulus

Once a design point is established, the secant modulus (E_s) is the slope of a line from the origin through the design point, Fig. 6.8.

f. Maximum Permissible Strain Data

6.2 Material Property Data Needed for Analysis   171

(a) For a low number (~1 - 10) of assembly/disassembly cycles

(b) For a higher number of assembly/disassembly cycles

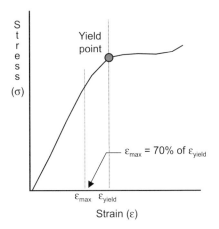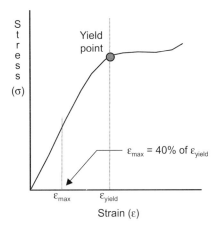

Figure 6.6 Design points for variable strain applications having a definite yield point

Values for maximum permissible strain of some groups and families of materials are given in Table 6.2. These can be useful for estimating initial performance but they should not be used for final analysis.

(a) For a low number (~1 - 10) of assembly/disassembly cycles

(b) For a higher number of assembly/disassembly cycles

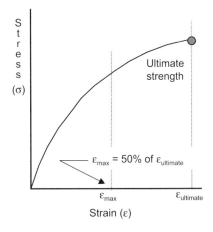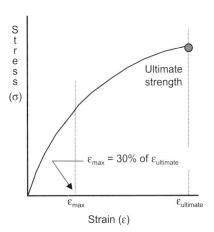

Figure 6.7 Design points for variable strain applications without a definite yield point

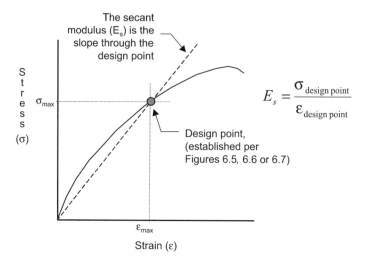

Figure 6.8 Calculating the secant modulus from the design point

This concludes the discussion of stress and strain. Important points to remember include:

- Published data in brochures is acceptable for initial screening but material data sheets and, preferably, actual stress-strain curves should be used to establish the design points for final analysis.
- Use stress-strain data that represents actual application conditions.
- Permissible strain tends to be higher for ductile and lower for brittle polymers.
- Recognize that many conditions may affect the actual maximum permissible strain and that end-use testing is necessary to verify predicted performance.

6.2.5 Coefficient of Friction (μ)

Coefficient of friction relates the normal force acting on an interface to the force required to slide one of the interface members across the other. It is used when calculating assembly or separation forces and retention strength, all situations where one snap-fit feature must slide across another.

Coefficient of friction is related to the lubricity of a material. Lubricity is the load bearing capability of the material under relative motion. It is a measure of the material's ability to slide across another material or itself without galling or other surface damage. Materials with good lubricity will tend to have lower coefficients of friction. Those with poor lubricity will tend to have higher coefficients of friction. Information on a material's lubricity can sometimes be found on its data sheet.

Some published coefficient of friction values are shown in Table 6.3. However, be aware that published values are based on specific tests and materials that may have little or no

Table 6.2 Maximum Permissible Strain

Material	Typical ε_{max}	Source
Most plastics fall within	1–10%	X
Glass filled plastics tend to fall within	1–2%	X
Polypropylene PP	8–10%	X
Polycarbonate 30% glass-fiber reinforced PC	1.8%	X
Polyphenylenesulfide (40% glass-fiber reinforced) PPS	1%	X
High heat polycarbonate PC	4%	B
Polycarbonate/ABS blend	2.5%	B
Acrylonitrile-styrene-acrylate ASA	1.9%	B
Polycarbonate blends	3.5%	B
Polycarbonate PC	4%	B
Polyamide (conditioned) PA	6%	B
Polyamide (dry) PA	4%	B
Polyamide/ABS	3.4%	B
Acrylonitrile-butadiene-styrene ABS	1.8%	B
Polycarbonate (10% glass reinforced)	2.2%	B
Polyamide/ABS (15% glass reinforced)	2.2%	B
Polycarbonate (20% glass reinforced)	2%	B
Polyamide conditioned (30% glass reinforced)	2%	B
Polyamide dry (30% glass reinforced)	1.5%	B
Polyetherimide PEI	9.8%	A
Polycarbonate PC	4–9.2%	A
Acetal	1.5%	A
Nylon 6 (dry)	8%	A
Nylon 6 (30% glass reinforced)	2.1%	A
Polybutylene terephthalate PBT	8.8%	A
Polycarbonate/Polyethylene terephthalate PC/PET	5.8%	A
Acrylonitrile-butadiene-styrene ABS	6–7%	A
Polyethylene terephthalate PET (30% glass reinforced)	1.5%	A

B—*Snap-fit Joints for Plastics = a design guide*, Polymers Division, Bayer Corp., 1998.
A—*Modulus Snap-Fit Design Manual*, Allied Signal Plastics, 1997.
X—Unidentified

- Materials in the table are unreinforced unless noted otherwise.
- These values are for short-term strain and low cycle or single cycle operation. For multiple cycles, use 60% of the values shown.
- The strain data is at room temperature.
- "Conditioned" refers to standard test conditions of 50% relative humidity and 20 °C, unless other specific humidity/temperature conditions are noted.
- "Dry" means low or no moisture content. Often it is "dry as molded".

relation to a specific application or to the common snap-fit condition of an edge sliding over a retention feature surface. The best source of friction data is testing under actual conditions, but this is rare. Use the published data along with lubricity information and your own

Table 6.3 Published Coefficients of Friction

Material	μ	Source	Notes
Polyetherimide PEI	0.20–0.25	A	*
Polycarbonate PC	0.25–0.30	A	*
Acetal	0.20–0.35	A	*
Nylon 6	0.17–0.26	A	*
Polybutylene terephthalate PBT	0.35–0.40	A	*
Polycarbonate/Polyethylene terephthalate PC/PET	0.40–0.50	A	*
Acrylonitrile-butadiene-styrene ABS	0.50–0.60	A	*
Polyethylene terephthalate PET	0.18–0.25	A	*
Polytetrafluoroethylene PTFE	0.12–0.22	B	**
Polyethylene PE rigid	0.20–0.25 (2.0)	B	**
Polypropylene PP	0.25–0.30 (1.5)	B	**
Polyaxymethelene; Polyformaldehyde POM	0.20–0.35 (1.5)	B	**
Polyamide PA	0.30–0.40 (1.5)	B	**
Polybutylene terephthalate PBT	0.35–0.40	B	**
Polystyrene PS	0.40–0.50 (1.2)	B	**
Styrene acrylonitrile SAN	0.45–0.55	B	**
Polycarbonate PC	0.45–0.55 (1.2)	B	**
Polymethyl methacrylate PMMA	0.50–0.60 (1.2)	B	**
Acrylonitrile-butadiene-styrene ABS	0.50–0.65 (1.2)	B	**
Polyethylene PE flexible	0.55–0.60 (1.2)	B	**
Polyvinyl chloride PVC	0.55–0.60 (1.0)	B	**
Slider specimen vs. Plate specimen	*At 10.6 mm/sec.*	T	***
Polypropylene (as molded) vs. Polypropylene (as molded)	0.71	T	***
Nylon (as molded) vs. Nylon (as molded)	0.65	T	***
Polypropylene (abraded) vs. Polypropylene (abraded)	0.27	T	***
Nylon (machined) vs. Nylon (machined)	0.47	T	***
Mild Steel vs. Polypropylene (abraded)	0.31	T	***
Mild Steel vs. Nylon (machined)	0.30	T	***
Polypropylene (abraded) vs. Mild steel	0.38	T	***
Nylon (machined) vs. Mild steel	0.40	T	***

A—*Modulus Snap-Fit Design Manual*, Allied Signal Plastics, 1997.
B—*Snap-fit Joints for Plastics a Design Guide*, Polymers Division, Bayer Corporation, 1998.
T—*Plastic Process Engineering*, James L. Throne, Marcel Dekker, Inc., 1979.
　* The values are for the given material tested against itself.
 ** Values are for the material tested against steel. Friction between different plastics will be equal to or slightly lower than these values. Friction between the same materials will generally be higher; a multiplier is shown in parenthesis if it is known.
*** Unlubricated tests, dynamic coefficient of friction.

judgment to determine a coefficient of friction. From the data shown, one can see that values of μ range from 0.2 to 0.7. For initial analysis, unless other information is available, values of 0.2 for low friction materials and 0.4 for high friction materials are reasonable estimates.

Coefficient of friction variability has a strong effect on the reliability and accuracy of assembly and retention calculations.

The data in Table 6.3 from [4] was associated with information on spin-welding and could have been developed with that technology in mind. However, it is useful in that it shows the kind of variation that can occur depending on the test. Note the difference between steel vs. polypropylene and polypropylene vs. steel, for example. All the coefficient of friction data should be considered by the designer as information that will allow for an "educated estimate" of the friction at the lock pair interface.

6.2.6 Other Effects

Plastic materials have many other properties that, while they do not appear in the calculations, can influence analysis because of their effect on stress and strain behavior. Some will also affect the dimensional stability of the parts.

Additives are chemicals added to enhance certain functional or processing capabilities of a plastic. Because additives may adversely affect mechanical properties, they can affect snap-fit feature performance. Examples of additives include impact modifiers, UV stabilizers, coloring agents and flame-retardants.

Plastics will exhibit *accelerated aging* at elevated temperatures. All plastics will experience degradation of mechanical properties at elevated temperatures over the long term. A comparison of thermal stability values will indicate the severity of the degradation. Sometimes stress-strain curves are generated to show performance at elevated temperatures.

Creep is a relatively long-term increase in strain (i.e. deflection) under a sustained load. The rate of creep for a material depends on the applied stress, temperature and time. Stress-strain curves showing the effects of long-term creep are required for long-term performance analysis. From these curves, a creep modulus can be determined and used in the calculations.

Plastic properties are sensitive to *temperature effects*. In general, materials become softer and more ductile and the modulus decreases with increasing temperature. The deflection temperature under load (DTUL), also called the heat deflection temperature or HDT, is a single point measurement that may be useful for quality control or for initial screening of materials for short-term heat resistance. However, the DTUL value should not be used as design data.

Fatigue endurance. For applications subjected to cyclic loads, SN curves can be generated. Cyclic loading, particularly reversing loads, can significantly reduce the life of a plastic part.

Notch sensitivity is the ease with which a crack propagates through a material from a notch, initial crack or a corner. A stress concentration factor related to the effect of sharp corners on local stress should be included in all calculations.

Chemical and ultra-violet effects may degrade mechanical properties. In general, as temperature and/or stress level increases, the plastic's resistance to these other effects will decrease.

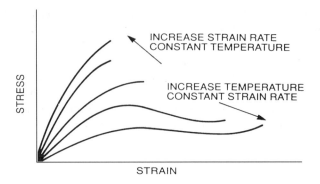

Figure 6.9 Effects of temperature and strain rate on stress-strain behavior, Courtesy of Ticona LLC, *Designing With Plastic—the Fundamentals*

Mold design and part processing can affect feature performance. Thick sections and improper cooling can cause voids or internal stresses. Mold flow patterns, knitlines and placement of gates can adversely affect feature strength.

Plastic behavior is *rate dependent*. This means it is affected by the speed of the applied load. Stress-strain tests are conducted at a standard speed and may not represent actual load rate in an application. For a given plastic, a high load rate will typically result in behavior similar to that at a low temperature: more rigid and brittle. A slow load rate results in behaviors similar to high temperature behavior (more ductile and flexible), Fig. 6.9.

The amount of recycled content or *regrind* as well as the effectiveness of the material mixing process (for uniformity) prior to molding can affect mechanical properties and part-to-part consistency.

Stress relaxation is a relatively long-term decrease in stress under a constant strain. (Creep involves constant stress; stress relaxation involves constant strain.) Data similar to creep data can be generated and a relaxation modulus determined, but relaxation data is not as available as creep data. The creep modulus can be used as an approximation of the relaxation modulus.

Toughness is the ability to absorb mechanical energy (impact) through elastic or plastic deformation without fracturing. Material toughness is measured by the area under the stress-strain curve. Tests for impact resistance under specific conditions include the Izod and Charpy tests of notched specimens, the tensile impact test and the falling dart impact test.

Water absorption. Some plastics, nylons for example, are very susceptible to moisture and humidity levels. Moisture content can affect mechanical properties as well as dimensional stability. Materials with low water absorption have better dimensional stability. Mechanical properties are often given at two humidity conditions: Dry as molded (DAM) and 50% relative humidity. Moisture content can affect mechanical properties (especially stiffness), electrical conductivity and dimensional stability. Nylon is particularly susceptible, use impact modified nylon to minimize moisture sensitivity.

Coefficient of Linear Thermal Expansion (CLTE) is a measure of the material's linear dimensional change under temperature changes. The lower the CLTE, the greater the dimensional stability. The mating and base parts should have similar values of CLTE if

Table 6.4 Published Coefficients of Linear Thermal Expansion (CLTE)

Material	in./in./°F 10^{-5}	cm/cm/°C 10^{-5}
Liquid crystal (GR*)	0.3	0.6
Glass	0.4	0.7
Steel	0.6	1.1
Concrete	0.8	1.4
Copper	0.9	1.6
Bronze	1.0	1.8
Brass	1.0	1.8
Aluminum	1.2	2.2
Polycarbonate (GR)	1.2	2.2
Nylon (GR)	1.3	2.3
TP polyester (GR)	1.4	2.5
Magnesium	1.4	2.5
Zinc	1.7	3.1
ABS (GR)	1.7	3.1
Polypropylene (GR)	1.8	3.2
Epoxy (GR)	2.0	3.6
Polyphenylene sulfide	2.0	3.6
Acetal (GR)	2.2	4.0
Epoxy	3.0	5.4
Polycarbonate	3.6	6.5
Acrylic	3.8	6.8
ABS	4.0	7.2
Nylon	4.5	8.1
Acetal	4.8	8.5
Polypropylene	4.8	8.6
TP polyester	6.9	12.4
Polyethylene	7.2	13.0

Courtesy of Ticona LLC, *Designing With Plastic—the Fundamentals*.
Also see [11] for additional CLTE data.
* GR indicates a glass-reinforced material.

possible. Careful consideration of constraint and compliance during feature selection will minimize the effects of CLTE differentials. Table 6.4 shows CLTE values for some plastics and, for comparison, some common metals.

Use CLTE to estimate compliance requirements in the interface, particularly when parts are large or differences between the expansion rates of the joined materials are significant. When these conditions exist, it is also more important to avoid over-constraint due to opposing features in the interface.

Mold shrinkage. Percentage of part shrinkage as it cools from the actual mold shape will affect final dimensions. In general, amorphous plastics have lower shrinkage than crystalline and glass-filled are lower than unfilled (neat) plastics. An excellent source of tolerance data for a wide variety of polymers is reference [5].

6.3 Cantilever Hook Design Rules of Thumb

The rules that follow are generally true but material, part and processing variation will affect their suitability for any given application. They can be useful for setting some nominal feature dimensions and providing a starting point for analysis. By taking the materials properties and variables discussed in the preceding section into account, the designer will be able to bias these rules of thumb in the right direction for more accurate estimates of dimensions.

Some of these guidelines are related to processing capabilities and following them can help avoid marginal processing situations that may cause inconsistent feature performance. As always, feature performance, especially on critical applications, must be verified by analysis and end-use testing.

Reflecting its popularity, there are many rules of thumb for the cantilever hook lock but few for other lock types. The rules are presented here in a logical order for most hook development situations. However, the designer should always keep in mind that hook design is frequently an iterative process. Initial dimensions are likely to require adjustment to account for design decisions made later in the process. Refer to Fig. 6.10 for the terminology used in the rules that follow.

6.3.1 Beam Thickness at the Base

Because the parent component dimensions and characteristics are usually fixed, they are the first constraints on feature design. Thus, we will start where the hook meets the parent component. A beam may extend from a wall or surface in many ways, the most common are a 90° protrusion and in-plane.

If the beam protrudes from a wall, Fig. 6.11a, then the beam thickness at its base (T_b) should be about 50 to 60% of wall thickness. Beams thinner than 50% may have filling and flow problems. Beams thicker than 60% may have cooling problems at the base because of the thick section. This may, in turn, lead to high residual stresses and voids which will weaken the feature (at its point of highest stress) and sink marks which are unacceptable on an appearance surface.

If the beam is an extension of a wall, Fig. 6.11b, then (T_b) should be equal to the wall thickness. If the beam thickness must be less than the wall thickness, then a gradual change in thickness over a length of the beam (at a 1 : 3 rate) from the wall to the desired beam thickness should be used to avoid stress concentrations and mold filling problems.

6.3.2 Beam Length

The total cantilever hook length (L_t) is made up of beam length (L_b) and retention feature length (L_r), Fig. 6.12. These two are considered separately because when bending is

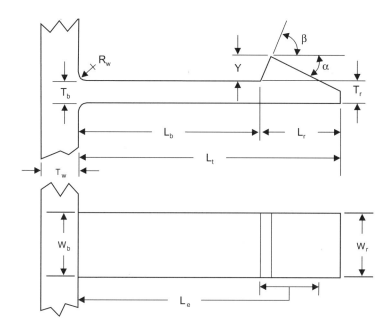

Dimensions shown in the hook drawings:
- L_r Retention feature length
- L_b Beam length
- L_t Total lock feature length
- T_w Wall thickness at the beam
- T_b Beam thickness at the wall
- T_r Beam thickness at the retention feature
- R_w Radius at the beam to wall intersection
- W_b Beam width at the wall
- W_r Beam width at the retention feature
- Y Undercut depth
- α Insertion face angle as designed (with the hook in its free state)
- β Retention face angle as designed (with the hook in its free state)

Other dimensions:
- δ Assembly deflection (Y and δ are typically equal)
- α_{max} Effective insertion face angle (hook at maximum assembly deflection)
- β_{min} Effective retention face angle (hook at maximum residual deflection)
- L_e Effective beam length (The distance from the base of the beam to the mating feature's point of contact on the insertion or retention face)

Figure 6.10   **Cantilever hook variables and terminology**

(a) Perpendicular to a wall (out of plane)

(b) In-plane from an edge

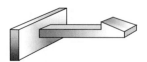

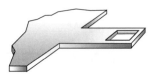

T_b should be 50% to 60% of the wall thickness

T_b should be equal to the edge thickness

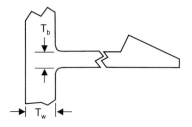

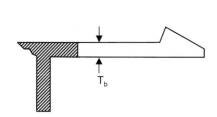

Figure 6.11 Initial beam thickness

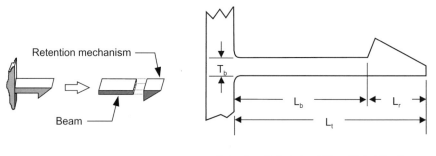

Beam length (L_b) should be at least 5 x T_b
A length of 10 x T_b is preferred

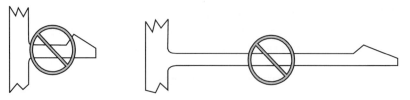

Figure 6.12 Initial beam length

calculated, only the flexible portion of the hook is considered. At this time, retention feature length is unknown, but we can establish the beam length now. Later, we can add retention feature length to beam length for total hook length. Ideally, we want to be free to select a beam length without any other limitations such as available space, but sometimes total length is limited by available space or mating part dimensions.

Beam length (L_b) should be at least 5× beam thickness ($5 \times T_b$) but 10× thickness ($10 \times T_b$) is preferred. Beams can be much longer than 10× thickness, but warpage and filling may become problems. Check the design against the material's spiral flow curves to ensure adequate feature filling.

Beams shorter than ($5 \times T_b$) will experience significant shear effects as well as bending at the base. Not only does this increase likelihood of damage during assembly, it renders the analytical calculations (based on beam theory) much less accurate. Shorter beams are much less flexible and create higher strains at the base. Longer beams are more flexible for assembly but also become weaker for retention.

Higher length to thickness ratios are recommended for plastics that are harder and more brittle.

6.3.3 Insertion Face Angle

Insertion face angle will affect the assembly force. The steeper the angle, the higher the force required to deflect and engage the hook.

Ideally, the maximum insertion face angle should be as low as possible for low assembly force, Fig. 6.13. An angle of 25–35° is reasonable. Angles of 45° or greater are difficult to assemble and should be avoided. For the common hook, the initial insertion face angle will also increase during insertion; another good reason to start out with that angle as low as possible. This change in angle that occurs during assembly is discussed in a later section.

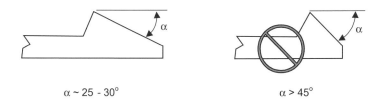

$\alpha \sim 25 - 30°$ $\alpha > 45°$

Figure 6.13 Initial insertion face angle

6.3.4 Retention Face Depth

The retention face depth (Y), sometimes called "undercut", determines how much the beam will deflect for engagement and separation, Fig. 6.14a. ("Separation" means both unintended release due to an external force or intentional release for disassembly.) For a

(a) Retention face depth

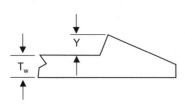

For $L_b/T_b \sim 5$, set $Y < T_w$
For $L_b/T_b \sim 10$, set $Y = T_w$

(b) Retention face angle

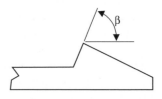

$\beta \sim 35°$ for a releasing lock with no external separation loads

$\beta \sim 45°$ for a releasing lock with low external separation loads

$\beta \sim 80 - 90°$ for a non-releasing lock with higher separation loads

Figure 6.14 Initial retention face depth and angle

beam length (L_b) to thickness (T_b) ratio in the range of 5:1, the initial retention face depth should be less than T_b. For a L_b/T_b ratio close to 10:1, the initial retention face depth can be equal to T_b. In general, for higher or lower ratios, Y should be adjusted accordingly. Harder and more rigid plastics can tolerate less deflection for a given length than can softer plastics.

Generally, for the beam/catch hook, the full retention face depth should be used for hook deflection and return. Thus retention face depth will equal deflection ($Y = \delta$). This helps ensure that separation forces on the catch occur as close as possible to the neutral axis of the beam and minimizes rotational forces at the end of the beam that would contribute to unintended release.

When analysis calculations are based on a known strain limit for the material, a maximum allowable deflection can be determined. The maximum retention face depth is then set equal to the maximum allowable deflection.

6.3.5 Retention Face Angle

Retention face angle will affect retention and separation behavior. The steeper the angle, the higher the retention strength and the disassembly force, Fig. 6.14b.

For a releasing lock where no external separation forces are acting on the mating part (aside from an intentional manual separation force) a retention face angle of about 35° is generally acceptable, Fig. 6.14c. The exact angle will depend on the coefficient of friction between the materials and the actual stiffness of the lock material. If the application is one with an expected high number of usage cycles (as with a moveable snap-fit) then a lower angle is preferred to reduce cyclic loading on both the lock and the mating feature. If the lock will be released only a limited number of times, then a higher angle may be possible.

If some relatively low external separation forces are expected, then a retention face angle of about 45° is a reasonable starting point. Again consider friction and hook stiffness. These locks may still be releasing, but manual separation forces will be high and a high number of removal cycles is not recommended.

If the lock must resist high external separation forces, then a releasing lock is not recommended and a permanent or non-releasing (manual deflection needed for disassembly) lock should be designed. The retention face angle should be close to 90°. A retention face angle of exactly 90° usually is not necessary. Because of frictional effects, any angle above a limiting value called the threshold angle will behave like a 90° angle.

6.3.6 The Threshold Angle

Because of friction between the feature contact surfaces, an angle less than 90° can still behave like a 90° angle. At an estimated coefficient of friction of 0.3, a good first approximation of this threshold angle is 80°. This means that any angle above 80° will behave like an angle of 90°. The threshold angle is a function of coefficient of friction and can be calculated if the coefficient of friction is known by solving the basic retention force formula for β as shown here:

$$\beta = \tan^{-1}\left(\frac{1}{\mu}\right) \quad (6.1)$$

Using an angle between the threshold angle and 90° on the retention face may sometimes be desirable because it will have slightly more dimensional compliance and robustness than a 90° angle can provide, Fig. 6.15.

6.3.7 Beam Thickness at the Retention Feature

Often the beam thickness at the retention face (T_r) is equal to the thickness at the feature base (T_b), Fig. 6.16a. However, when strains at the base are high, tapering the beam over its length will more evenly distribute strain through the beam and reduce the chances of overstrain at the base, Fig. 6.16b. Common taper ratios ($T_b : T_r$) range from 1.25 : 1 up to 2 : 1. In shorter beams, tapering can reduce strains at the base by as much as 60%. However, tapering will also reduce the retention strength. Tapering is one possible solution to high strains when design constraints force a beam to violate the 5 : 1 minimum length to thickness rule.

Do not taper a cantilever beam from the retention face to the base. This moves virtually all the strain to the base of the hook and damage is very likely.

(a) Retention face of 90° is less robust to dimensional variation

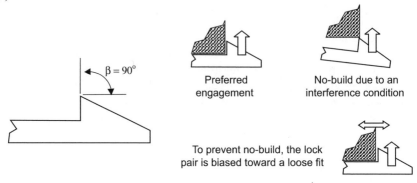

(b) Retention face of $90° \geq \beta \geq \beta_t$ is more robust

Figure 6.15 Retention face threshold angle

6.3.8 Beam Width

Most beams have a constant width from the base to the retention face. When this is the case, beam width does not affect the maximum assembly strain but it does affect assembly and disassembly forces and retention strength. Because strain is not a function of beam width when the width is constant, beam strength can be improved by increasing the width without causing an increase in the strain, Fig. 6.17a. This can be an alternative to increasing the beam thickness when more retention strength is needed. (Increasing beam thickness will cause an increase in strain.)

For beam theory to apply, the width should be less than or equal to the length, Fig. 6.17b. As the width becomes greater than 1/2 the length, the feature begins to behave more like a plate than a beam. However, given the other variables involved in the calculations, relatively minor inaccuracies at higher beam widths can generally be ignored.

Beams can be tapered on width, just as they can be tapered on thickness, Fig. 6.17c. Tapering the beam in the width dimension can reduce strain at the base, but not as effectively as tapering the thickness. Where a beam extends in-plane from an already thin wall, tapering the width may be the only option.

A beam must have a 4:1 taper in width to get the level of same strain reduction as a beam with a 2:1 taper in thickness.

(a) No taper, $T_b = T_r$

(b) A 2:1 taper ($T_b = 2 \times T_r$) is common

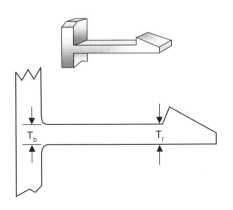
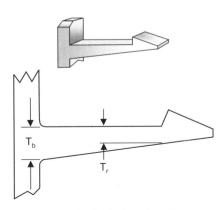

Improved strain distribution along the beam means lower strain at the wall, particularly in shorter beams

(c) Improper taper, $T_b < T_r$

Figure 6.16 Beam thickness at end, constant section and tapering

6.3.9 Other Features

The rules given here for the cantilever hook can sometimes be used to establish initial dimensions for other lock features. For example, rules for insertion and retention face angles are generally applicable to catch-like retention features used in torsional lock configurations. Rules for beam bending apply to trap lock behavior during assembly. For a loop style lock, the insertion and retention face angles are found on the mating catch and they do not change with assembly deflection or residual deflection as with the hook.

6.4 Initial Strain Evaluation

With these intial values for the hook dimensions, we now know beam deflection, thickness and length. If the beam width is constant, a quick preliminary calculation of the maximum assembly strain at the base can be made:

$$\varepsilon_{\text{initial}} = 1.5 \frac{T_b \delta}{L^2} \tag{6.2}$$

(a) No taper on width, $W_b = W_r$.

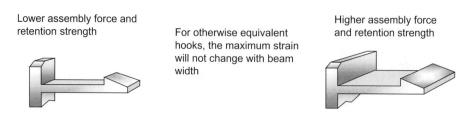

Lower assembly force and retention strength

For otherwise equivalent hooks, the maximum strain will not change with beam width

Higher assembly force and retention strength

(b) Effect of high beam width

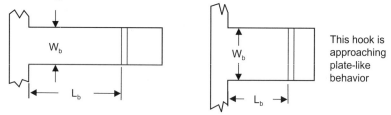

This hook is approaching plate-like behavior

As beam width to length ratio increases, behavior becomes less like a beam and more like a plate, the threshold is roughly $W_b > L_b$

(c) Width-tapered hooks

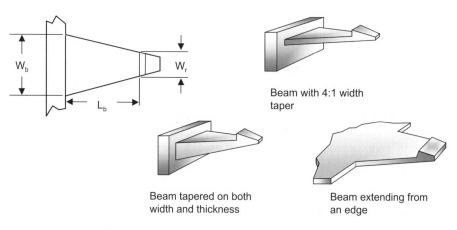

Beam with 4:1 width taper

Beam tapered on both width and thickness

Beam extending from an edge

Figure 6.17 Beam width

The result can be compared to the maximum allowable strain. (Remember, for evaluating assembly behavior, use the dynamic strain limit if it is available.) This early calculation will indicate if the proposed hook design is reasonably close to the maximum allowable strain. Do not worry if the design is over the allowable strain by as much as 50% at this time. Other conditions not yet considered will tend to reduce the actual strain in the final analysis. As one becomes familiar with how these conditions affect strain, you will develop a feeling for just how much effect these effects will have on the final strain numbers. The conditions include:

- Possible parent material deflection at the feature's base during assembly.
- Possible mating feature deflection during assembly.

On the other hand, if the calculation indicates the maximum strain far exceeds the permissible strain (more than two or three times greater, for example) then changes to the initial beam dimensions (thickness, length or retention face depth) can be made. A glance at the deflection magnification effects for the hook/wall configuration being designed will also give an indication of how much the calculated strain will be reduced and whether an adjustment is necessary at this time.

Again, keep in mind that the rules presented in this section are useful for establishing *initial* dimensions for a cantilever hook or other snap-fit feature. Detailed analysis and end use testing are still required to ensure feature performance meets all application requirements. To summarize some important earlier remarks:

- These are general rules of thumb and, in some cases, reflect preferred design practices for ease of molding.
- They can be useful for setting some nominal feature dimensions that provide a starting point for analysis.
- They can help avoid marginal processing situations that may cause inconsistent feature performance.
- Analyzing plastic features involves many variables and some assumptions. It is not an exact science.
- Feature design is normally an iterative process.
- Feature performance, especially on critical applications must be verified by analysis and end-use testing.

6.5 Adjustments to Calculations

Cantilever hook analysis is based on classic structural beam theory. Adjustments to the results are sometimes needed to reflect real part behavior. Before feature calculations are discussed in more detail, we will introduce several important adjustments to the basic beam calculations. They are:

- *Stress concentrations (k)*, where abrupt changes in section can cause an increase in local strain. Stress concentrations tend to increase the actual strain at the base of the beam so they reduce the maximum allowable calculated strain. Although called "stress concentrations", it is appropriate to apply the adjustment to strain in the calculations when we are not working with stress.
- *Wall deflection*, expressed as a deflection magnification factor (Q), where wall deformation tends to reduce the actual strain and increase the actual deflection of a hook under a given load. Deflection magnification will also tend to reduce retention strength.
- *Mating feature deflection* (δ_m), where some assembly deflection occurs in the other member of the lock pair. Mating feature deflection will tend to reduce assembly deflection and strain as well as separation strength.
- *Effective angle* (α_e and β_e), where assembly or separation deflection changes the insertion or retention face angles and affects the predicted performance of the lock. Effective angle does not affect strain, but it can have significant effects on assembly force. It can sometimes affect retention strength.

6.5.1 Adjustment for Stress Concentration

Stress concentrations occur where the feature undergoes a sudden change in section. The effect of stress concentrations is to increase the actual strain in the part above the strain calculated from beam theory. For hooks and beams in general, we are most concerned with the area stressed in tension where the feature meets a wall. A radius at that location will reduce stress concentration effects, but they cannot be totally eliminated.

Figure 6.18 shows a curve for the stress concentration factor (k) vs. the ratio of beam thickness to the radius at the beam to wall juncture. Other sources show curves similar to the one shown here. As shown on the graph, a value of $k = 1.5$ is reasonable. A stress concentration factor of 1.0 is impractical because the very large radius required would result in voids, residual stresses and sink marks. Rules for process-friendly design also limit this radius to about 50% of beam thickness.

The stress concentration factor is applied as shown here:

$$\varepsilon_{max} = \frac{\varepsilon_{design}}{k} \qquad (6.3)$$

Therefore our goal is:

$$\varepsilon_{calc} \leq \varepsilon_{max} \qquad (6.4)$$

There also seems to be a consensus in the literature of:

- A minimum radius of 0.5 mm (0.020 in.) permitted for stressed areas.
- A minimum radius of 0.13 mm (0.005 in.) permitted for unstressed areas.

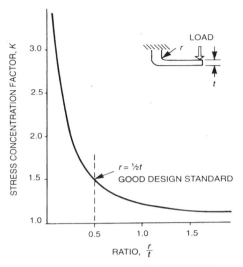

Figure 6.18 Stress concentration factor. Courtesy of Ticona LLC, *Designing With Plastic—the Fundamentals*

6.5.2 Adjustment for Wall Deflection

Beam calculations assume that the base area (i.e. the wall) from which a feature protrudes is infinitely stiff. In other words, no base deflection occurs as the feature deflects. In reality, base deflection can occur, Fig. 6.19, and the effect on feature behavior can be significant. Some sources [6, 7, 8, 9, 10] discuss these effects in great detail. The information in this section is adapted from [6]. At an intuitive level, the reader should understand that when base deflection occurs, the actual forces, strengths, stresses and strains in the beam are less than the calculated values.

Figure 6.20 shows how behaviors other than bending become more significant as the beam becomes shorter. As the beam length-to-thickness ratio becomes smaller, the base deflection (wall) effects in particular become significant [9]. The wall's elasticity and its effect on beam behavior is accounted for by the deflection magnification factor (Q) [6].

Including deflection magnification effects in calculations gives more accurate results in the form of:

- Lower strains and lower beam deflection force than predicted by beam theory.
- Lower installation force, retention strength and disassembly force than predicted by beam theory.
- Higher allowable beam deflection than predicted by beam theory.

(a) In all beam calculations, beam theory assumes all the deflection occurs in the beam

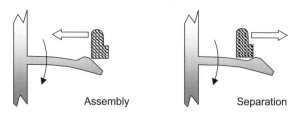

(b) In reality, deflection can also occur in the wall on which the beam is mounted

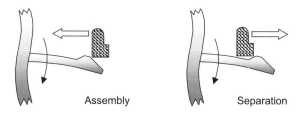

(c) At equivalent strains or deflection forces, the actual deflection is greater than predicted, thus the name "deflection *magnification*"

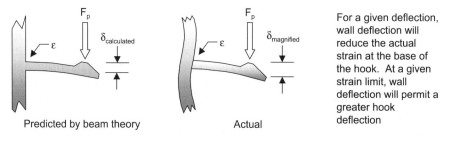

For a given deflection, wall deflection will reduce the actual strain at the base of the hook. At a given strain limit, wall deflection will permit a greater hook deflection

Figure 6.19 Wall effects on beam bending and deflection magnification

By contrast, ignoring deflection magnification effects can result in:

- Calculated strain at a given deflection is too high
- Allowable deflection at a given strain is too low
- Pessimistic results for assembly
- Optimistic results for retention

6.5 Adjustments to Calculations

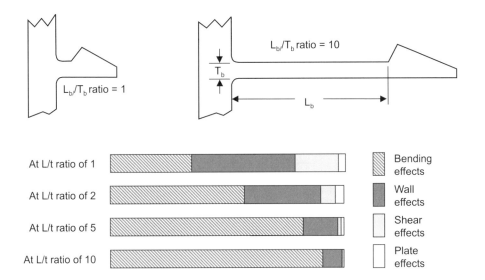

Figure 6.20 Effect of beam length on beam behavior

Tables 6.5 and 6.6 give values of Q for constant section beams and for beams having a thickness taper ratio of 2:1. The beam/wall configurations represented in these tables are shown in Figs. 6.21 and 6.22 respectively. The deflection magnification factor (Q) is used in the beam behavior calculations as:

$$\varepsilon = 1.5 \frac{T_b \delta}{L^2 Q} \tag{6.5}$$

$$F_P = \frac{W_b T_b^2 E \varepsilon}{6 L_b Q} \tag{6.6}$$

Use caution when using the deflection magnification factor. It is quite easy to improperly apply it by using it more than once in a series of calculations. For example, if the Q-factor is used to calculate a value for actual strain, do not use it again when using that actual strain value to calculate a deflection force. The correct deflection force will automatically result because the corrected value of strain is being used in the force calculation.

6.5.3 Adjustment for Mating Feature Deflection

Every constraint feature is part of a constraint pair. As feature performance is analyzed, remember that the mating feature and its parent component may also deflect. If this deflection is significant, it can have major effects on the calculations. Mating feature deflection affects assembly and disassembly forces and strains and retention strength in exactly the same manner as deflection magnification.

Table 6.5 Values of the Deflection Magnification Factor (Q) for a Constant Rectangular-Section Beam

Beam aspect ratio L_b/T_b	Beam to wall configuration (refer to Fig. 6.21)				
	1 Beam ⊥ to a solid wall	2 Beam ⊥ and in interior area of wall	3 Beam ⊥ to wall and parallel at edge	4 Beam ⊥ to wall and parallel at edge	5 Beam in-plane with wall at edge
1.5	1.60	2.12	2.40	6.50	8.00
2.0	1.35	1.70	1.90	4.60	5.50
2.5	1.22	1.45	1.65	3.50	4.00
3.0	1.17	1.35	1.45	2.82	3.15
3.5	1.15	1.28	1.38	2.40	2.65
4.0	1.14	1.25	1.36	2.25	2.40
4.5	1.13	1.23	1.33	2.10	2.20
5.0	1.12	1.21	1.28	1.95	2.10
5.5	1.11	1.19	1.27	1.85	1.95
6.0	1.10	1.17	1.25	1.75	1.85
6.5	1.09	1.15	1.24	1.70	1.80
7.0	1.08	1.13	1.22	1.65	1.75
7.5	1.07	1.11	1.20	1.60	1.70
8.0	1.06	1.10	1.19	1.55	1.65
8.5	1.05	1.09	1.18	1.50	1.60
9.0	1.04	1.08	1.17	1.45	1.57
9.5	1.03	1.07	1.16	1.40	1.55
10.0	1.02	1.06	1.16	1.38	1.52
10.5	1.01	1.05	1.15	1.36	1.50
11.0	1.00	1.04	1.15	1.35	1.47

Values interpreted from Q Factor graphs in the *Modulus Snap-Fit Design Manual*, Allied Signal Plastics, 1997.

Mating feature deflection can be measured or calculated [12]. A graphic solution to determining the effects of mating feature deflection is shown in Fig. 6.23.

First make a judgment: If the mating feature/part can be considered stiff relative to the subject feature, then mating part deflection need not be considered. If the mating feature/part is flexible, then its deflection effect must be determined. Plot both force/deflection curves using the same scales.

Calculate the lock feature force vs. deflection curve over the estimated range of lock feature deflection, Fig. 6.23a. (Be sure to include deflection magnification effects in these calculations.) Plot the deflection as shown. It is usually enough to know only two or three points to construct the deflection curve, unless the degree of curvature is high. (Recall the discussion of assembly force-deflection signature in Chapter 3.) The calculation for deflection force is described in an upcoming section.

6.5 Adjustments to Calculations

Table 6.6 Values of the Deflection Magnification Factor (Q) for a Rectangular-Section Beam with a 2 : 1 Taper

Beam aspect ratio L_b/T_b	Beam-wall configuration*		Beam aspect ratio L_b/T_b	Beam-wall configuration*	
	2T Beam ⊥ and in interior area of wall	5T Beam in-plane at edge		2T Beam ⊥ and in interior area of wall	5T Beam in-plane at edge
2.0	1.60	3.50	7.0	1.14	1.52
2.5	1.50	3.00	7.5	1.13	1.47
3.0	1.40	2.50	8.0	1.13	1.43
3.5	1.33	2.25	8.5	1.12	1.40
4.0	1.25	2.05	9.0	1.12	1.38
4.5	1.22	1.90	9.5	1.11	1.35
5.0	1.20	1.80	10.0	1.11	1.32
5.5	1.17	1.70	10.5	1.10	1.30
6.0	1.15	1.65	11.0	1.10	1.28
6.5	1.14	1.58	—	—	—

Values interpreted from Q Factor graphs in the *Modulus Snap-Fit Design Manual*, Allied Signal Plastics, 1997.
* Refer to Fig. 6.22.

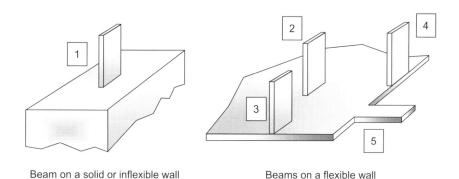

Beam on a solid or inflexible wall   Beams on a flexible wall

Figure 6.21 Beam to wall configurations for constant section rectangular beam (for use with Table 6.5), adapted from [6]

Calculate or measure mating feature/part deflection vs. force over the expected range of locking feature deflection. Plot the deflection as shown. Note that the mating feature deflection is negative relative to the lock's deflection. Again, it is usually sufficient to have only two or three points to construct the curve, Fig. 6.23b.

When these curves are superimposed, the intersection is the actual lock feature deflection and deflection force, Fig. 6.23c. Use these values as you proceed with the analysis. A pure

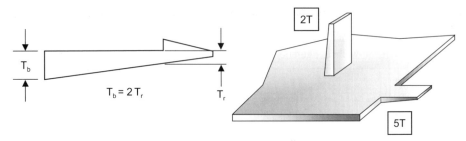

Tapered beams on a flexible wall.

Figure 6.22 Beam to wall configurations for rectangular section beam with a 2 : 1 taper (for use with Table 6.6), adapted from [6]

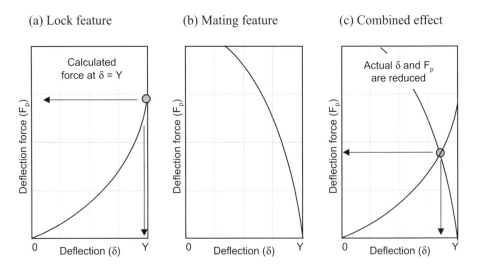

Figure 6.23 Graphical solution for the effect of mating feature deflection, adapted from [12]

mathematical solution (finding the intersection by solving for the common solution to the two deflection equations) is also possible but generally not worth the trouble. The graphical method is relatively simple and effective.

Note that the retention face depth (Y) should still be set equal to the total deflection because it must accommodate all the deflection in the system, not just the lock feature deflection. For calculation purposes, the lock feature deflection is now δ_{max}. This is an instance where the undercut (Y) does not equal the lock's actual deflection.

6.5.4 Adjustment for Effective Angle

Most published calculations for cantilever hook behavior do not make clear the effect of hook deflection on the insertion face angle (α) and retention face angle (β). They typically

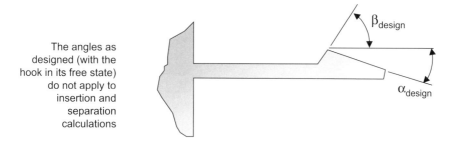

The angles as designed (with the hook in its free state) do not apply to insertion and separation calculations

Figure 6.24 The retention and insertion face angles as designed

show sample calculations using values of α or β for the hook "as designed" or in its free state, Fig. 6.24. In reality, the insertion face angle can change significantly as the hook deflects with significant effects on the calculated assembly force. Changes in the retention face angle are less dramatic, but do occur and can sometimes affect the retention strength. The actual angles must be adjusted to reflect an *effective angle* of the insertion and retention faces. If changes in the insertion and retention face angles are ignored:

- Calculated assembly force will be lower than actual.
- Calculated retention strength *may* be higher than actual.

Some effects related to insertion and retention face angles were introduced in Chapters 3 and 4 when assembly and retention signatures were discussed.

6.5.4.1 Effective Angle for the Insertion Face

The maximum insertion face angle occurs at maximum assembly deflection. Therefore, to calculate the maximum assembly force, we must determine the angle at that point, Fig. 6.25. Neglecting beam curvature and hook end rotation, a simple calculation for the change in insertion face angle is:

$$\Delta \alpha = \tan^{-1}\left(\frac{\delta_{max}}{L_e}\right) \tag{6.7}$$

This calculation assumes no hook end rotation and no beam curvature during deflection. When a beam is very long in relation to its thickness or when a beam is tapered, rotation may be significant. However, this simplified calculation will bring the calculated force much closer to actual than just ignoring the change in insertion face angle altogether. A more complex calculation that takes beam curvature and end rotation is possible but normally not necessary. Once the change in angle is known, it is added to the design angle to give the effective insertion face angle (α_{max}):

$$\alpha_{effective} = \alpha_{max} = \alpha_{design} + \Delta \alpha \tag{6.8}$$

This value for maximum insertion face angle should be used in all assembly force calculations.

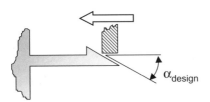
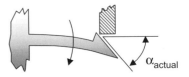

The design angle applies only when the mating feature first engages the hook

As the mating feature moves up the insertion face, the angle α increases

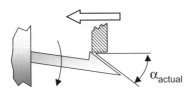
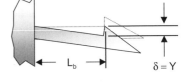
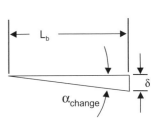

Neglecting beam curvature and end rotation simplifies the calculation

The change in angle is calculated from deflection and beam length

Figure 6.25 Effective angle for the insertion face

6.5.4.2 Effective Angle for the Retention Face

For the retention face, we are concerned with the effective angle causing a *lower* retention force than expected. The minimum retention face angle occurs when the hook moves to its final locked position. If tolerances or misalignment cause it to retain some residual deflection, the retention face angle will be reduced. Normally, this effect is relatively insignificant, but, depending on the application, it may become important. To calculate the minimum retention strength, we must determine the angle at the maximum possible residual deflection, Fig. 6.26. Neglecting beam curvature and any hook end rotation, the same simple calculation used for the change in insertion face angle can be applied using maximum residual deflection as the deflection variable:

$$\Delta \beta = \tan^{-1}\left(\frac{\delta_{\text{residual}}}{L_e}\right) \tag{6.9}$$

6.5 Adjustments to Calculations

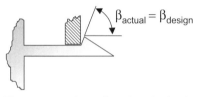
The design angle applies when the hook returns to its original position after assembly

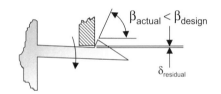
If some deflection remains in the hook, then the retention face angle is reduced

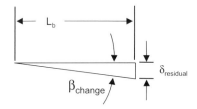
As with the insertion face angle, neglecting beam curvature and end rotation simplifies the calculation

Figure 6.26 Change in retention face angle with residual deflection

The calculated change in angle is *subtracted* from the original design angle to give the effective retention face angle:

$$\beta_{\text{effective}} = \beta_{\text{design}} - \Delta\beta \qquad (6.10)$$

This value should be used in all retention strength calculations.

6.5.5 Adjustments Summary

These adjustments will be applied to the hook analysis calculations that follow. Their effects are summarized in Table 6.7.

Table 6.7 Summary of Adjustments to Calculated Strain

	Effect on actual strain	Effect on assembly force	Effect on separation force
Stress concentration (k)	increase	—	—
Deflection magnification (Q)	reduce	reduce	reduce
Mating feature deflection (δ_m)	reduce	reduce	reduce
Effective insertion face angle (α_{actual})	—	increase	—
Effective retention face angle (β_{actual})	—	—	reduce

6.6 Assumptions for Analysis

In addition to the assumptions about material properties discussed earlier in this chapter, we must make certain assumptions about the hook so that the classic beam equations can be applied:

- The beam material is homogeneous with the same modulus of elasticity in tension as in compression.
- The beam is straight or has a curvature in the plane of bending with a radius of curvature at least 10 times the beam depth.
- The beam cross-section is uniform.
- The beam has at least one longitudinal plane of symmetry.
- All loads and reactions are perpendicular to the beam's axis, and lie in the same plane, which is the longitudinal axis of symmetry.
- The beam is long in proportion to its depth.
- The beam is not disproportionately wide.
- The maximum stress does not exceed the proportional limit.
- Applied loads are not impact loads.

Calculations for plastic materials are immediately subject to error because of plastic's visco-elastic behavior. When specific applications violate other assumptions, accuracy of the results is even more questionable. Keep this in mind when setting design targets and safety factors for your features. The more assumptions violated, the less representative of actual feature behavior the calculations will become.

6.7 Using Finite Element Analysis

When too many materials or analytical assumptions are violated, consider using finite-element analysis if the application merits it. Find a detailed discussion of finite element analysis for snap-fit features in reference [11]. Consider FEA when:

- Complex beam shapes or sections must be analyzed.
- Complex stress/strain conditions exist.
- Deflections are large.
- Too many assumptions are violated.
- Plate-like deflections occur (the beam is wide relative to its length).

Remember that proper constraint in the attachment is always a requirement. While finite element analysis is capable of analyzing improperly constrained attachments, the attachment itself is fundamentally incorrect and likely to have problems.

6.8 Determine the Conditions for Analysis

A complete analysis will involve extensive data. The kind of information needed for a complete analysis includes:

- The range of plastic material properties for both new and for aged parts. (Developed from statistical treatment of raw data if possible.)
- Typical mold tolerances for the feature material [5] should be used to estimate all mating and base part maximum and minimum material conditions that affect feature performance.
- Coefficient of linear thermal expansion (CLTE) for the mating materials.
- Temperature history for the application.
- Intended application usage (function):
  1 cycle of use (assembly only)
  Limited cycles (maintenance or service) usually 3–10 cycles
  Multiple cycles (moveable attachment) $\gg$ 10 cycles.

Determine worst case combinations of conditions and material properties for analysis as appropriate. Depending on availability of data and the application, a complete analysis under all conditions may not be necessary or possible.

6.9 Cantilever Hook Analysis for a Constant Rectangular Section Beam

A simple cantilever hook is a constant cross-section rectangular beam with a catch retention feature at the end. It is the most common style of lock feature, (although we have seen that it is far from the most effective or efficient). The hook variable names introduced in Fig. 6.10, are repeated in Fig. 6.27 for reference during the following discussion. After this description of the analysis procedure for a cantilever hook, the analytical procedures for tapered beams are given.

6.9.1 Section Properties and the Relation between Stress and Strain

While we generally do not need to consider stress in our calculations, some formulae related to stress are shown here for reference.

As we saw in the stress-strain curves and the calculation for secant modulus, stress and strain are related through the modulus of elasticity (E):

$$E = \frac{\text{stress}}{\text{strain}} = \frac{\sigma}{\varepsilon} \qquad (6.11)$$

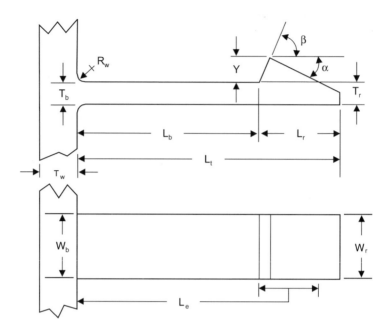

Dimensions shown in the hook drawings:

L_r Retention feature length
L_b Beam length
L_t Total lock feature length
T_w Wall thickness at the beam
T_b Beam thickness at the wall
T_r Beam thickness at the retention feature
R_w Radius at the beam to wall intersection
W_b Beam width at the wall
W_r Beam width at the retention feature
Y Undercut depth
α Insertion face angle as designed (with the hook in its free state)
β Retention face angle as designed (with the hook in its free state)

Other dimensions:

δ Assembly deflection (Y and δ are typically equal)
α_{max} Effective insertion face angle (hook at maximum assembly deflection)
β_{min} Effective retention face angle (hook at maximum residual deflection)
L_e Effective beam length (The distance from the base of the beam to the mating feature's point of contact on the insertion or retention face)

Figure 6.27 Cantilever hook variables and terminology

6.9 Cantilever Hook Analysis for a Constant Rectangular Section Beam

Maximum bending stress is at the beam surface farthest from the neutral axis and farthest from the point of maximum beam deflection and we generally care about maximum tensile stress rather than compression stress. Stress is in units of MPa, (newtons per mm^2).

Beams with rectangular sections are by far the most common hook configuration and are used in all the examples here. Formulae for calculating the properties of other common sections can be found in many structural engineering references. The ones shown here were found in references [2, 12]. For a beam having a rectangular section, the section properties are calculated as:

$$I = \frac{\text{base} \times \text{height}^3}{12} \quad (6.12)$$

$$Z = \frac{\text{base} \times \text{height}^2}{6} \quad (6.13)$$

Stress for a rectangular section beam is:

$$\sigma = \frac{Mc}{I} = \frac{F_p L_b}{Z} \quad (6.14)$$

The deflection at the end of a cantilevered beam is:

$$\delta = \frac{F_p L_b^3}{3EI} \quad (6.15)$$

I is the section moment of inertia and is in units of mm^4; c is the distance of the outer surface from the neutral axis in mm. In a rectangular section, c is one-half the beam thickness. The outer surface is where the highest tensile and compressive stresses occur. Usually we care about tensile stress because it is responsible for the strains that cause hook damage and failure. Z is the section modulus in units of mm^3.

6.9.2 Evaluating Maximum Strain

Figure 6.28 shows an example application. The initial dimensions for this application were determined using the rules of thumb given in Section 6.3. The calculations are applied using this example as we step through the process.

It is common to begin calculations with a given deflection and solve for strain. If the calculations are manual, it is desirable to begin with initial hook dimensions that are as close as possible to final in order to simplify the work.

This is less important when software for analysis is available and many design alternatives can be evaluated quickly. However, when using software for beam analysis, be aware that many of the available beam analysis packages do not comprehend all of the four calculation adjustments discussed in Section 6.5. For more accurate results, these adjustments must be used to fine-tune the results of software-based calculations. Because we are interested in the highest possible strain in the hook, select dimensions for the calculation that reflect maximum material conditions for beam thickness, undercut and mating feature interference.

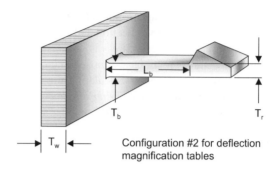

Configuration #2 for deflection magnification tables

Dimensional information:

Beam length (L_b) = 15 mm

Wall thickness at the beam (T_w) = 4 mm

Beam thickness at the wall (T_b) = 2 mm

Beam thickness at the retention feature (T_r) = 2 mm

Radius at the beam to wall intersection (R_w) = 1 mm

Beam width at the wall (W_b) = 3 mm

Beam width at the retention feature (W_r) = 3 mm

Undercut depth ($\delta = Y$) = 2 mm

Residual deflection (δ_r) = 0.1mm

Insertion face angle as designed (α) = 25°

Retention face angle as designed (β) = 50°

Material information:

E_s = 2000 MPa

Strain at design point (ε_{max}) = 3%

Coefficient of friction (μ) ~ 0.4

Figure 6.28   Hook data for example calculations

Because the primary design constraint is usually the material's maximum allowable dynamic strain (the design point), begin the calculations with strain as the limiting variable. The process of determining the maximum allowable strain and the design point has already been discussed.

6.9.2.1   Adjusting Maximum Allowable Strain for Stress Concentrations

The stress concentration factor (k) can be applied in one of two ways: to reduce the maximum allowable strain or to increase the calculated strain ($\varepsilon_{calculated}$) which will then be compared to the maximum allowable strain (ε_{design}). This is valid because:

$$k\varepsilon_{calc} \leq \varepsilon_{design} \tag{6.16}$$

is equivalent to:

$$\varepsilon_{calc} \leq \frac{\varepsilon_{design}}{k} \quad \text{and} \quad \varepsilon_{calc} \leq \varepsilon_{max} \tag{6.17}$$

6.9 Cantilever Hook Analysis for a Constant Rectangular Section Beam

We will choose to use the stress concentration factor to reduce the maximum allowable strain because the calculated strain will be adjusted by other factors and can also be used as the basis for other calculations. If calculated strain is modified by k, and then used in later calculations these calculations may be incorrect. From Fig. 6.18, a stress concentration factor of $k = 1.5$ is found for the R_w/T_w ratio of 0.5. This value is used to reduce the maximum allowable strain for the example application according to Equation 6.3:

$$\varepsilon_{max} = \frac{\varepsilon_{design}}{k} \quad \text{so} \quad \varepsilon_{max} = \frac{0.03}{1.5} \quad \text{and} \quad \varepsilon_{max} = 0.02 \tag{6.18}$$

6.9.2.2 Calculating the Maximum Applied Strain in a Constant Section Beam

We are, of course, most concerned with maximum tensile strain. The maximum strain in the deflected beam will occur at the intersection of the beam to the wall on the tensile side of the neutral axis. The strain is calculated as:

$$\varepsilon = 1.5 \frac{T_b \delta}{L^2} \tag{6.19}$$

As applied to the example:

$$\varepsilon_{initial} = 1.5 \frac{2 \times 2}{15^2} \quad \text{and} \quad \varepsilon_{initial} = 0.027 = 2.67\% \tag{6.20}$$

A maximum allowable strain (ε_{max}) of 2% is used for the example. Comparing this initial calculated value $\varepsilon_{initial}$ of 2.7% to ε_{max} of 2.0% we note that although the calculated strain is higher, it is reasonably close. Knowing that several adjustments to this value are yet to come, we do not yet make any changes to the design.

Keep in mind that the strain value we have just calculated is at a given deflection. This deflection (δ) is equal to the height (Y) of the catch.

6.9.2.3 Adjusting the Calculated Strain for Deflection Magnification

Recall the discussion of deflection magnification. Any deflection of the wall or surface on which the hook is mounted will reduce the actual strain at the base of the beam. The calculated strain should now be reduced accordingly. The Q factor is found in Table 6.5 for beam/wall configuration #2 with a L_b/T_b ratio of 15/2 or 7.5. The value of Q is 1.11 and this value is used to recalculate the strain as:

$$\varepsilon_{calc} = 1.5 \frac{T_b \delta}{L^2 Q} \tag{6.21}$$

For the example application:

$$\varepsilon_{calc} = 1.5 \frac{2 \times 2}{15^2 \times 1.11} = \frac{\varepsilon_{initial}}{Q} \quad \text{and} \quad \varepsilon_{calc} = 0.024 = 2.4\% \tag{6.22}$$

Caution: A common mistake is to apply the deflection magnification adjustment more than once in a series of hook calculations. It is critical to understand that once the adjustment for deflection magnification has been made, any following calculation that uses an adjusted strain value must not be modified a second time by the Q factor. This is illustrated in the next section with respect to the force calculation.

6.9.3 Calculating Deflection Force

Knowing the strain (e_{initial}) and the hook's dimensions and section properties, we can now calculate the deflection force. To calculate deflection force *using hook dimensions and e_{initial}* the deflection magnification factor should be used (again, from Table 6.5) and the basic calculation is:

$$F_P = \frac{W_b T_b^2 E \varepsilon}{6 L_b Q} \tag{6.23}$$

Note that, as when adjusting the strain for wall deflection, the Q factor is used to reduce the value of F_P. For the example:

$$F_P = \frac{3 \times 2^2 \times 2000 \times 0.0267}{6 \times 15 \times 1.11} \quad \text{and} \quad F_P = 6.4 \text{N} \tag{6.24}$$

If we calculate deflection force *using $\varepsilon_{\text{calc}}$*, the deflection magnification factor is NOT used because the strain has already been adjusted for deflection magnification. The basic calculation would again be:

$$F_P = \frac{W_b T_b^2 E \varepsilon}{6 L_b} \tag{6.25}$$

For the example:

$$F_P = \frac{3 \times 2^2 \times 2000 \times 0.024}{6 \times 15} \quad \text{and} \quad F_P = 6.4 \text{N} \tag{6.26}$$

We see that, for the example application, the calculated values of the deflection force in both Equation 6.24 and Equation 6.26 are equal. Note how the wrong answer would be obtained in Equation 6.26 had the deflection magnification factor been used in that calculation.

The deflection force that has been calculated is the force to bend the beam and we have assumed that it is applied at the end of the beam. In reality, the deflection force is applied at the point of contact as the mating feature moves across the insertion or retention faces of the catch at the end of the beam. Generally, this variation is ignored but if, for a specific application, it is determined to be important, appropriate adjustments to the calculations can be made and the effective beam length (L_e) would be used.

6.9.4 Adjusting for Mating Part/Feature Deflection

We now know force and deflection values for the hook. If the mating component is stiff relative to the feature, no mating part deflection adjustment is needed. Otherwise, we must account for mating part deflection.

First plot the lock feature's force deflection curve, Fig. 6.29. For simplicity of the example, we will assume a straight line relationship so all that is needed is one pair of force and deflection values (F_p and $d_{\max}$). F_p has just been calculated and $d_{\max}$ is known from the initial hook dimensions.

Determine the mating part's force/deflection curve over the same range of deflection. This may require additional calculations or physical measurement. Plot this curve on the same graph, Fig. 6.29. For this example, we will assume a deflection for the feature on the mating part has already been determined and it is also shown on the graph.

We see that the total deflection required for engagement is actually shared by the feature and the mating part. Actual feature deflection and deflection force are less than originally calculated. This means the original strain results can be adjusted.

We can take the actual deflection as found above divided by the original design deflection ($\delta = Y$) and get a factor (f) for adjusting some of the original values.

$$f = \frac{\delta_{\text{actual}}}{\delta_{\text{design}}} = \frac{1.48}{2.0} = 0.74 \tag{6.27}$$

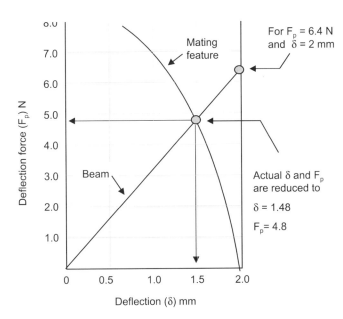

Figure 6.29 Solving for the effects of mating feature deflection for the example hook application

Since strain and deflection are directly related, use this factor to reduce the calculated actual strain for the example:

$$\varepsilon_{final} = 0.74 \times 0.024 = 0.0178 = 1.78\% \tag{6.28}$$

Because we began these calculations with the design point strain and most of our adjustments to the calculations have tended to reduce strain, we may now find that our calculated maximum actual strain in the beam is considerably less than the maximum allowable strain as indicated by the design point (and adjusted for stress concentrations).

For the example, we find that the final calculated strain is indeed below the allowable maximum strain of 2.0%.

With the margin provided by this new value of strain, we may wish to make some changes in the feature. Remember, however, assembly and separation forces have not yet been calculated and it may be advisable to wait until all performance values are known. If changes are made, redo the calculations with the new numbers and plot these results against the mating feature deflection to verify their acceptability.

6.9.6 Determine Maximum Assembly Force

Maximum assembly force is important because we must verify that assembly forces do not violate ergonomic rules for maximum forces applied by fingers, thumbs or hands.

Use the maximum value for beam bending force after all deflection effects are taken into account. Find the coefficient of friction from Table 6.4 or from supplier data. If coefficient of friction data is not available, estimate it at around 0.2 to 0.4 depending on the lubricity of the material(s), surface roughness and a bias toward identifying high or low force depending on the application. Maximum assembly force is found by the calculation:

$$F_{assembly} = F_P \frac{\mu_{dynamic} + \tan \alpha_{effective}}{1 - (\mu_{dynamic} \tan \alpha_{effective})} \tag{6.29}$$

However, the value of the insertion face angle (α) used in this formula must first be adjusted for beam deflection.

6.9.6.1 Determine the Effective Insertion Face Angle

Effective angle was introduced in Section 6.5.4. Unless the insertion face angle does not change during engagement (as with a loop engaging a catch), a modified value for α (α_{actual}) is needed for the assembly force calculation. The simplified calculation for the change in angle was given in Equation 6.7. Add the change in the insertion face angle to the original angle (Equation 6.8) and use the resulting value of α_{max} when calculating the maximum assembly force. For the example application:

$$\Delta\alpha = \tan^{-1}\left(\frac{\delta}{L_e}\right) \quad \text{so} \quad \Delta\alpha = \tan^{-1}\left(\frac{2}{15}\right) \quad \text{and} \quad \Delta\alpha = 7.6° \tag{6.30}$$

$$\alpha_{effective} = \alpha_{design} + \delta\alpha \quad \text{so} \quad \alpha_{effective} = 25° + 7.6° \quad \text{and} \quad \alpha_{effective} = 33° \tag{6.31}$$

Now applying Equation 6.29:

$$F_{assembly} = F_P \frac{\mu_{dynamic} + \tan \alpha_{effective}}{1 - (\mu_{dynamic} \tan \alpha_{effective})} \quad \text{so} \quad F_{assembly} = 4.8 \frac{0.4 + 0.65}{1 - (0.4 \times 0.65)}$$

$$\text{and} \quad F_{assembly} = 6.8 \text{N} \tag{6.32}$$

6.9.7 Determine Release Behavior

Release behavior has several meanings depending on whether we are talking about intentional or unintentional release. They are calculated the same way, but if we are concerned with ease of separation (in a releasing attachment) we calculate a maximum value. If we are concerned with retention strength, we would calculate a minimum value.

Separation force (F_s) is the effort required for a person to separate the parts. For a releasing lock, disassembly involves applying a force in the separation direction to one of the parts. For a non-releasing lock (requiring manual deflection), disassembly involves applying a bending force directly to the lock to unlatch it from the mating feature. With the lock deflected, the parts can be separated. When calculating disassembly force, we calculate a maximum value.

Retention strength (F_r) is the lock's resistance to unintended release. When we calculate retention strength, we calculate a minimum value because minimum retention strength must be greater than any forces in the part separation direction. Retention strength calculations must assume one of several possible failure modes. For cantilever hooks, these are:

- *Bending*, where the mating feature slides over the retention face, the hook bends and releases. For a releasing lock, this is the common retention behavior.
- *Shear*, where some portion of the constraint pair fails in shear. Shear calculations are simply based on the applicable cross-sectional area and the shear strength of the material. Because of their simplicity, they are not discussed here.
- *Tension*, where some portion of the constraint pair fails in tension. These calculations are based on the applicable cross-sectional area and an appropriate tensile strength limit (yield, maximum or ultimate) of the material. Like shear calculations, because of their simplicity, they are not discussed here.
- *Combination*, which is a complex set of effects where some combination of bending, shear, tension and retention mechanism rotation cause distortion and release. Calculations of this behavior are beyond the scope of this chapter and normally beyond the capability of simple hand calculations. They may require finite element analysis.

For permanent and non-releasing locks, shear or a combination of retention behaviors as discussed above are more likely. For a releasing lock where release is determined by bending, the behavior is calculated in a manner similar to the assembly behavior.

For brevity, the term *retention strength* is used in the following discussion although, depending on the application, we may actually be interested in separation force.

If the retention face angle changes due to residual deflection in the hook, an adjustment to the design angle is needed if the effect is determined to be significant. This adjustment results in an effective retention face angle (β_{actual}). The simplified formula (Equation 6.9) is the same as that for the change in insertion face angle, except that residual deflection is used in the calculation.

For the example application:

$$\Delta\beta = \tan^{-1}\left(\frac{\delta_{residual}}{L_e}\right) \quad \text{so} \quad \Delta\beta = \tan^{-1}\left(\frac{0.1}{15}\right) \quad \text{and} \quad \Delta\beta = 0.4° \quad (6.33)$$

This calculated change in retention face angle is relatively insignificant but, for the sake of the example, we will subtract it from the design value of β as shown in Equation 6.10:

$$\beta_{effective} = \beta_{design} - \Delta\beta \quad \text{so} \quad \beta_{effective} = 50° - 0.4° \quad \text{and} \quad \beta_{effective} = 49.6° \quad (6.34)$$

The calculation for separation force is similar to that for assembly force:

$$F_{separation} = F_P \frac{\mu_{static} + \tan\beta_{effective}}{1 - (\mu_{static}\tan\beta_{effective})} \quad \text{so} \quad F_{separation} = 4.8\frac{0.4 + 1.17}{1 - (0.4 \times 1.17)}$$
$$\text{and} \quad F_{separation} = 14.2\text{N} \quad (6.35)$$

For the example hook, we have calculated its performance as:

- Maximum assembly strain (ε_{final}) = 1.78%
- Assembly deflection (δ) = 1.48 mm
- Deflection force (F_p) = 4.8 N
- Maximum assembly force ($F_{assembly}$) = 6.8 N
- Minimum separation force ($F_{separation}$) = 14.2 N

6.9.7.1 Additional Retention Considerations

If release involves the same deflection and behaviors as assembly (bending along the same beam axis, for example), then the maximum allowable strain is already known. If release is quick, as for a releasing lock, then the comparison to the dynamic strain limit made for assembly strain will still apply.

If disassembly involves (slower) manual deflection, then several factors may change the maximum strain calculation and must be considered. First, the manual deflection (d_{manual}) necessary for release may result in greater deflection than assembly simply because hook movement is not based on a fixed physical attribute (like Y). Use the maximum possible manual deflection to calculate strain. The enhancements called guards can help limit manual deflection if necessary.

A second effect is the longer-term deflection that can occur during manual deflection. Strain levels that may be acceptable when compared to the dynamic strain limit may not be safe when compared to a static strain limit.

Evaluating retention behavior may also require evaluating possible damage to the hook during a loading cycle. For a non-releasing hook for example, a high load on the parts in the separation direction may cause combined bending and tensile stresses at the lock's base or combined tensile, bending and shear stresses at the retention mechanism. Methods for combining these stresses exist and are described in structural mechanics books.

6.10 Cantilever Hook Tapered in Thickness

Tapered beams, Fig. 6.30, offer an advantage over straight beams in stress/strain and assembly force reduction. (A possible disadvantage is the reduction in retention strength.) Tapering the beam thickness is more effective than tapering beam width and is preferred. Beams can generally be tapered anywhere from 1.25 : 1 up to 2 : 1. The shorter the beam, the greater the impact of tapering on strain reduction.

The procedures for determining strength, forces, stresses and strains for tapered beams are identical to those for constant section beams. However, the strain calculation is different. The applicable strain calculation for a thickness-tapered beam is:

$$\varepsilon_{\text{calc}} = 1.5 \frac{T_b \delta}{L^2 QK} \qquad (6.36)$$

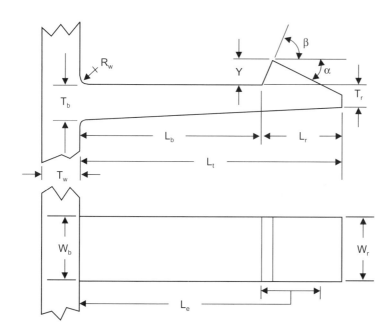

T_b Beam thickness at the wall $T_b > T_r$
T_r Beam thickness at the retention feature T_b / T_r is the (thickness) taper ratio

Figure 6.30   **The thickness-tapered beam**

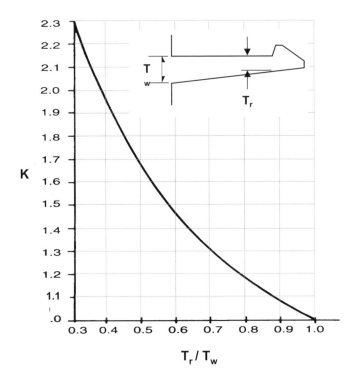

Figure 6.31 The proportionality constant (K) for thickness tapered beams, Adapted from Ticona LLC, *Designing With Plastic—the Fundamentals*

Where K is found in Fig. 6.31 as a function of the ratio (T_r/T_b). Use this value of strain to proceed through the remainder of the calculations as described above. Values for the deflection magnification factor (Q) for beams with a 2:1 thickness taper are given in Table 6.7.

When calculating deflection force from the strain adjusted for deflection magnification, the formula for a tapered beam is the same as that for the constant section beam (Equation 6.25):

$$F_P = \frac{W_b T_b^2 E \varepsilon}{6 L_b} \tag{6.37}$$

Although any taper ratio is possible, a 2:1 taper is common. If a 2:1 taper is applied to the example application so that $(T_r = 1)$ and all other dimensions remain the same, we find $K = 1.67$ from Figure 6.31 and $Q = 1.13$ from Table 6.6. and the calculations are:

$$\varepsilon_{calc} = 1.5 \frac{T_b \delta}{L^2 Q K} \quad \text{so} \quad \varepsilon_{calc} = 1.5 \frac{2 \times 2}{15^2 \times 1.13 \times 1.67} \quad \text{and} \quad \varepsilon_{calc} = 1.4\% \tag{6.38}$$

$$F_P = \frac{W_b T_b^2 E \varepsilon}{6 L_b} \quad \text{so} \quad F_P = \frac{3 \times 2^2 \times 2000 \times 0.014}{6 \times 15} \quad \text{and} \quad F_P = 3.7 \text{N} \tag{6.39}$$

Note the significant reductions in these strain and force values from the (non-tapered) beam used in the preceding example. Proceed with the assembly force calculations as described above for the constant section beam.

6.11 Cantilever Hook Tapered in Width

When beam thickness is limited, (possibly because the beam is an in-plane extension of a wall) tapering on width is an option, Fig. 6.32. Tapering the beam width is less effective than tapering beam thickness because thickness in the bending force equations is a second order term while the beam width is a first order term. (A 4 : 1 taper in width is required in order to have the same effect as a 2 : 1 thickness taper.) Again, the shorter the beam, the greater the effect of tapering on strain reduction.

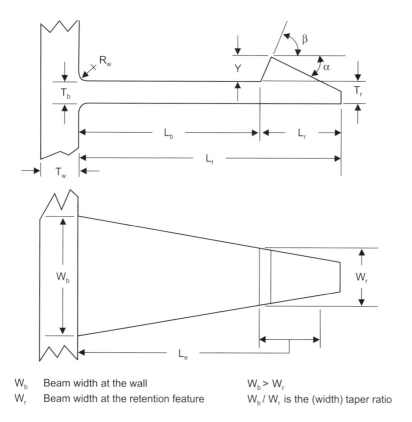

W_b Beam width at the wall
W_r Beam width at the retention feature

$W_b > W_r$
W_b / W_r is the (width) taper ratio

Figure 6.32 **The width-tapered beam**

Although any taper ratio is possible, a 4:1 taper is common. Values for a deflection magnification factor for beams tapered on width are not provided, but a reasonable approximation for the deflection magnification can be made by selecting the appropriate beam/wall configuration and choosing an aspect ratio that will create an equivalent bending moment at the wall.

The strain calculation for a 4:1 width-tapered beam is:

$$\varepsilon_{\text{calc}} = 1.17 \frac{T_b \delta}{L^2 Q} \tag{6.40}$$

When calculating the deflection force from the strain, the formula for a width-tapered beam is the same as that for the constant section beam (Equation 6.23):

$$F_P = \frac{W_b T_b^2 E \varepsilon}{6 L_b} \tag{6.41}$$

Proceed with the adjustments and assembly force calculations as described above. If a deflection magnification factor is to be applied, it would be used as shown.

6.12 Cantilever Hook Tapered in Thickness and Width

Sometimes it is desirable to taper beams on both thickness and width, Fig. 6.33. Again, the only difference is that the calculations become more complex. A discussion of computing the behavior of these beams can be found in [11].

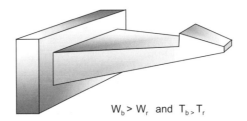

Figure 6.33 **Beam tapered in both thickness and width**

6.13 Modifications to the Insertion Face Profile

As discussed in Chapter 3, the insertion face profile can be modified to improve the insertion force-time signature. The profile can be determined by calculating the instantaneous angle at several points on the insertion face and then constructing the insertion face profile as a curve tangent to these angles. The calculation is based on the simplified calculation for the change in insertion face angle (Equation 6.7). As shown in Fig. 6.34:

$$\Delta \alpha = \tan^{-1}\left(\frac{\delta}{L_e}\right) \qquad (6.42)$$

$$\alpha_{design} = \alpha_0 - \Delta \alpha \qquad (6.43)$$

6.14 Modifications to the Retention Face Profile

Chapter 4 also introduced the concept of a more desirable retention face profile for improved retention performance. In a manner similar to that for the insertion face profile, the retention face profile is determined by calculating the instantaneous angle at several points on the retention face and then constructing the profile as a curve tangent to these angles. The calculation is based on the estimate of the change in retention face angle (Equation 6.9). As shown in Fig. 6.35:

$$\Delta \beta = \tan^{-1}\left(\frac{\delta}{L_e}\right) \qquad (6.44)$$

$$\beta_{design} = \beta_0 + \Delta \beta \qquad (6.45)$$

6.15 Other Feature Calculations

The living hinge was defined (in Chapter 4) as a locator feature. Calculations for living hinge behavior can be found in [11]. Some sources of information for calculating the behavior of other lock feature styles are listed in Table 6.8.

6.16 Summary

This chapter provided a brief overview of some important materials issues related to feature strength and analysis. Because many excellent sources of information exist on the subject of feature level calculations, the chapter simply provides an overview of the primary calculations involved in hook analysis. Rules of thumb for establishing initial feature dimensions were given. Following these rules should provide a reasonable hook design as a starting point for analysis.

(a) For a constant insertion face angle (α_{actual}), calculate α_{design} at the point of contact

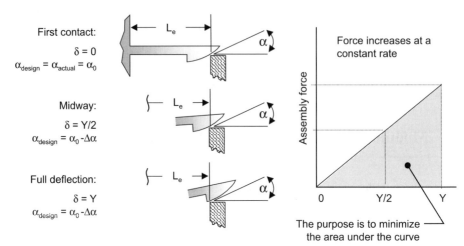

First contact:
$\delta = 0$
$\alpha_{design} = \alpha_{actual} = \alpha_0$

Midway:
$\delta = Y/2$
$\alpha_{design} = \alpha_0 - \Delta\alpha$

Full deflection:
$\delta = Y$
$\alpha_{design} = \alpha_0 - \Delta\alpha$

Force increases at a constant rate

The purpose is to minimize the area under the curve

(b) Calculating the adjustment

$$\Delta\alpha = \tan^{-1}\frac{\delta}{L_e}$$

$\Delta\alpha$ is calculated from instantaneous deflection and effective beam length

(c) For an over-center effect

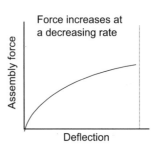

Add additional degrees to the calculated Δa

$$\alpha = \alpha_0 - (\Delta\alpha + \alpha_{additional})$$

Force increases at a decreasing rate

Figure 6.34 Designing an insertion face profile

6.16 Summary   215

(a) With a flat retention face, the signature may be concave, flat or convex

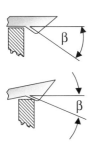 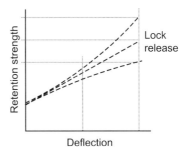

Instantaneous retention strength is a function of increasing deflection force and a decreasing angle β

$$\beta_{actual} = \beta_0 - \Delta\beta$$

(b) To ensure β_{actual} remains constant for maximum retention strength and maximum energy absorption, the design angle β_{design} must be adjusted by $\Delta\beta$

At full engagement:

$$\delta = 0$$
$$\beta_{design} = \beta_{actual} = \beta_0$$

Midway to release:

$$\delta = Y/2$$
$$\beta_{design} = \beta_0 + \Delta\beta$$

Full deflection:

$$\delta = Y$$
$$\beta_{design} = \beta_0 + \Delta\beta$$

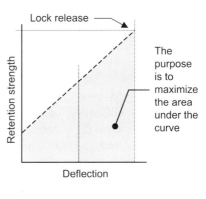

The purpose is to maximize the area under the curve

(c) Calculating the adjustment

$$\Delta\beta = \tan^{-1}\frac{\delta}{L_e}$$

$\Delta\beta$ is calculated from instantaneous deflection and effective beam length

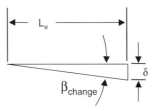

Figure 6.35 Designing a retention face profile

Table 6.8 Sources of Calculation Information for Other Lock Features and Shapes

Annular locks	*Snap-fit Joints for Plastics—A Design Guide*
Beams with complex sections	Bayer Corporation
Torsional locks	Pittsburgh, PA
Beams with complex sections	*Designing with Plastics—The Fundamentals*
Varieties of cantilever conditions	Design Manual TDM-1, 1996. Ticona LLC Summit, NJ
L-shaped beams	*Modulus Snap-Fit Design Manual*
U-shaped beams	Allied Signal Inc. Morristown, NJ
Annular locks	*Designing Plastic Parts for Assembly*
Beam tapered on length and width	Paul A. Tres
Closed-form beam calculations	Hanser/Gardner Publications, Inc
Finite element analysis	Cincinnati, OH
Living hinges	
Torsional locks	

All of these sources also contain information about calculations for constant rectangular section beams. This list is not all inclusive, many other resin suppliers provide design information about snap-fits, including: BASF, Dow Plastics, DuPont, GE Plastics and Monsanto.

Most importantly, some modifications to the basic calculations were described in this chapter. These modifications should be used to adjust the results of the basic beam calculations described in the literature. The designer should be careful to understand which adjustments are already included and which must be applied when using any design formulae from the literature or using analysis software.

6.16.1 Important Points in Chapter 6

- Use material property data from product brochures and sales literature only for initial screening and rough estimates of performance.
- Material data sheets can provide more application specific data and more complete data than brochures. Use this data form for more accurate calculations for initial evaluation and design. Recognize that many conditions may affect the actual maximum permissible strain and that end-use testing is necessary to verify predicted performance.
- Actual stress-strain curves are the preferred source for stress-strain data.
- Use stress-strain data that represents actual application conditions whenever possible, but no matter how representative the data is with respect to the application, end-use testing is the only way to verify feature performance.
- Maximum allowable strain tends to be higher for ductile polymers and lower for brittle materials.
- Tapering the beam can significantly reduce strain at the base.

- In a constant width hook, the strain is independent of beam width. Retention strength can be increased by increasing the width with no increase in strain.

References

1. CAMPUS® [Computer Aided Material Preselection by Uniform Standards] is a registered trademark of Chemie Wirtschaftsforderungs-Gesellschaft (CWFG). It is distributed free of charge to qualified customers. CAMPUS website: www.campusplastics.com
2. *Designing with Plastics—The Fundamentals, Design Manual TDM-1*, (1996) Ticona LLC. Summit, NJ (Formerly Hoechst Celanese Corporation, now a division of Celanese AG.)
3. *Plastic Part Design for Injection Molding*, 1994, Robert A. Malloy, Hanser/Gardner Publications, Inc., Cincinnati OH.
4. Plastics Process Engineering, 1979, James L. Throne. Marcel Dekker, Inc. New York.
5. *Standards and Practices of Plastics Molders*, 1998 Edition. Molders Division of The Society of the Plastics Industry, Inc. Washington, D.C.
6. *Modulus Snap-Fit Design Manual*, 1997, Allied Signal Plastics, Morristown, NJ.
7. *New Snap-Fit Design Guide*, 1987, Allied Signal Plastics, Society of Plastics Engineers ANTEC, 1987.
8. *Improving Snap-Fit Design*, 1987, C.S. Lee, A. Duban, E. D. Jones, Plastics Design Forum, Sept./Oct. 1987.
9. *Snap-Fit Design*, July 1977, W.W. Chow, University of Illinois, Urbana, Department of Mechanical Engineering.
10. *Parametric Investigations of Integrated Plastic Snap Fastener Design*, 1994, P. Kar, J. Renaud, University of Notre Dame, Proceedings of S. M. Wu Symposium on Manufacturing Science at Northwestern University.
11. Designing Plastic Parts for Assembly—Paul A. Tres, Hanser/Gardner Publications, Inc., Cincinnati OH, 2000.
12. *Snap-Fit Joints for Plastics—a design guide*, 1998, Polymers Division of the Bayer Corporation, Pittsburgh, PA.

Bibliography

Automated Program for Designing Snap-Fits—G. G. Trantina and M. D. Minnicbelli, GE Plastics, Pittsfield, MA, Plastics Engineering, August 1987.
Beyond the Data Sheet—Designer's guide to the interpretation of data sheet properties, David R. Rackowitz, BASF Plastic Materials, Wyandotte, MI.
Designing Cantilever Snap-Fit Latches for Functionality—Technical Publication #SR-402, Borg-Warner Chemicals.
It's a SNAP!, Zan Smith, Hoechst Celanese Corporation, Summit NJ, Assembly Magazine, October 1994.
Snap-Finger Design Analytics and Its Element Stiffness Matrices—Dhirendra C. Roy, United Technologies Automotive, SAE Technical Paper Series (SP-1012), International Congress and Exposition, 1994.
Standard Test Method for Kinetic Coefficient of Friction of Plastic Solids, ASTM Standard D 3028, ASTM Committee D-20 on Plastics.
The Give and Take of Plastic Springs, Z. Smith, M. Fletcher, D. Sopka, pp. 69–72, Machine Design, November 1997.
Understanding Tight-Tolerance Design, R. Noller, pp 61–72, Plastics Design Forum, March/April 1990.

7 The Snap-Fit Development Process

The purpose of the snap-fit development process is to: Produce an attachment between components of defined *basic shapes*, for an application requiring a certain locking *function*, using *constraint* and *enhancement* features in an interface between the mating part and base part brought together in a selected *engage direction* using a particular *assembly motion*.

This chapter describes the process by which the elements, key requirements and snap-fit concepts that were discussed in Chapters 2 through 4 are brought together to create a fundamentally sound snap-fit application. The chapter begins with a brief explanation of the rational behind the snap-fit development process followed by a step-by-step discussion of the process. The reader should understand that this is an idealized process and the realities of a product-engineering project may force modifications to it. The core principles of the process, however, should always apply. These important principles are identified as they appear in the discussion.

This chapter discusses only the development process itself. The details of feature analysis and problem diagnosis are discussed in Chapters 6 and 8 respectively.

In Chapter 1, five important skills for snap-fit development were introduced:

- Knowledge (of snap-fit technology and design options)
- Spatial reasoning
- Attention to detail
- Creativity
- Communication.

The development process will enable the designer to apply all of these skills while creating the snap-fit attachment. The reader will also find that much is made of manual activities like hand drawing a concept sketch, handling parts during benchmarking and making crude representative models of the parts under development. These activities are critical parts of the spatial reasoning and creative aspects of the development process and they should not be regarded as unimportant and ignored. "...the hand speaks to the brain as surely as the brain speaks to the hand." [1]

7.1 Introduction

This section is an explanation of the reasoning behind the attachment level development process for snap-fits. It is useful background information but if the reader wishes to skip this section, the discussion of the process itself begins at Section 7.2.

Note the difference between development and design. As the terms are used here, development means the entire process of conceptualizing, creating, designing and testing a snap-fit application. Development includes design. Design is the development process step where feature dimensions and tolerances are established and detailed drawings are made. This step often includes analysis, but many times, the initial feature dimensions are determined by following past experience or general rules of thumb. Analysis is applied only if indicated by prototype testing.

7.1.1 Concept Development vs. Detailed Design

The reader will notice that considerable effort (Steps 1 through 3) is spent on developing the concept during the snap-fit development process, Fig. 7.1. Actual snap-fit feature design does not begin until Step 4. One might be tempted to ask, "Why should I spend so much time on the concept? Why can't I jump into design right away?"

Studies [2, 3] have shown that as much as 70 to 80% of a product's total cost to produce are established during the concept stage of product development. In other words, if you do not do a good job creating the product concept, you have already locked yourself into a more expensive product design. Other studies have shown that changes made later in the development process become much more expensive and, once tooling has been made or the parts are in production, the cost to make changes (improvements) is often prohibitive [4, p. 128]. Other studies note the high "leverage" one has over the product in the concept stage in terms of quality and the ability to implement changes [5]. In other words, the concept stage can make or break an application in terms of both cost and quality. This is a basic tenant of design for assembly and is true of the attachment as well as the product as a whole. The attachment level process begins in the concept stage of product development when the designer can have significant impact on the product. By the time the parts get to the feature analysis and design step, they will be both fundamentally sound and cost effective. We will begin by briefly explaining the rationale behind the snap-fit development process.

7.1.2 A General Development Process

A basic attachment development process can be described as having two stages, Fig. 7.2. In the first, an attachment idea or concept is generated. In the second, the attachment concept is analyzed and designed.

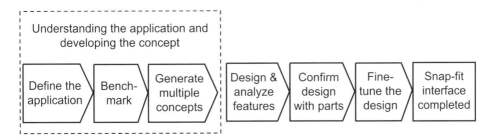

Figure 7.1 The snap-fit development process

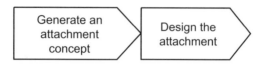

Figure 7.2 A general development process

Often, the tendency during product development is to adopt an existing attachment concept and jump quickly to design, Fig. 7.3. This is attractive because it is fast and has a certain amount of security in knowing the attachment has been used before. (Although one may not know if it worked well.) Copying or simply modifying what has been done before, however, will prevent the designer from considering other possible attachment options. It can also leave one open to repeating others' mistakes. This can result in poor attachments or costly re-engineering when prototype parts reveal shortcomings. On the other hand, if an entirely new concept is developed, the designer runs the risk of venturing into uncharted territory. In many cases, regardless of the approach used, only one attachment strategy is considered with little effort spent developing alternatives that may be better than the first idea. Limitations of both knowledge and time contribute to this situation.

To ensure the final attachment is the best it can be, a better approach is to develop several concepts following a structured thinking process. Both existing and new ideas can then be combined to produce the best of both worlds. Thus, the basic process that was shown in Fig. 7.2 is expanded to the more desirable process shown in Fig. 7.4. This improved process makes the desirable approach of developing several concepts more explicit by dividing the original "generate an attachment concept" step into two steps: "develop alternative concepts" followed by "evaluate alternatives and select the best concept".

The develop attachment concepts step in this improved process is the creative "heart" of the snap-fit development process. However, to jump immediately into creativity without preparation or follow-up can be counterproductive at best, disastrous at worst. A more desirable approach is *controlled creativity*, in which knowledge about the application and attachment level principles are focused to drive creative solutions that are also practical. Therefore, preparation and follow-up steps are added to complete a preferred development process for snap-fits, Fig. 7.5.

Next, by considering the attachment level elements, (the spatial and descriptive "objects" one must consider when developing a snap-fit) we can adapt the preferred process to one that is specific to snap-fits, Fig. 7.6.

The attachment level development process is not in conflict or disagreement with other product development processes. Let us compare it to another process particularly appropriate

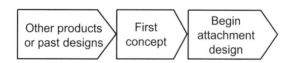

Figure 7.3 Typical snap-fit development process

7.1 Introduction 221

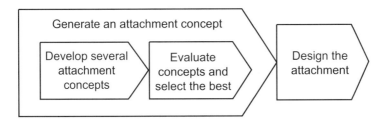

Figure 7.4 Improved development process

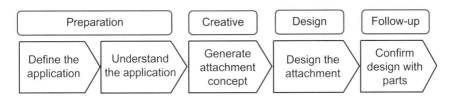

Figure 7.5 Preferred development process

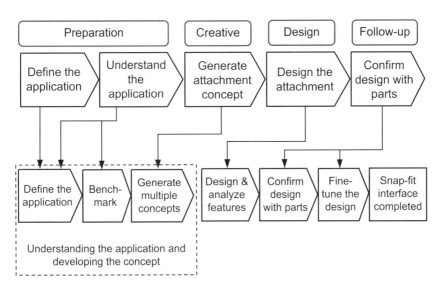

Figure 7.6 The generic preferred process leads to the snap-fit development process

to plastic parts and snap-fit development. Malloy [4, p. 130] describes such a process as having these nine steps:

1. Defining end-use requirements.
2. Create preliminary concept sketch.
3. Initial materials selection.
4. Design part in accordance with material properties.
5. Final materials selection.
6. Modify design for manufacturing.
7. Prototyping.
8. Tooling.
9. Production.

We can then show in Fig. 7.7 how the major steps of the snap-fit development process map to Malloy's process. Note: The Malloy book is an extremely valuable reference with regard to overall plastic part design as the complex issues around each of the above nine steps are discussed in great detail.

This introduction has explained why the snap-fit development process looks like it does. Next, we will describe the tasks and decisions associated with each step of the process. The key requirements and the elements of the Attachment Level Construct were introduced in Chapter 2 and discussed in detail in Chapters 3 and 4. We will now learn how they are used during the snap-fit development process. For reference, the model of the entire construct (introduced in Chapter 1) is repeated in Fig. 7.8. The relationship of the snap-fit elements to the development process is shown in Fig. 7.9.

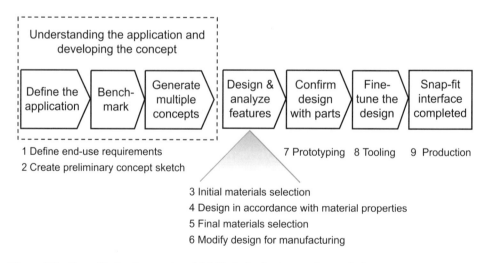

Figure 7.7 Snap-fit development and Malloy's basic stages of part design

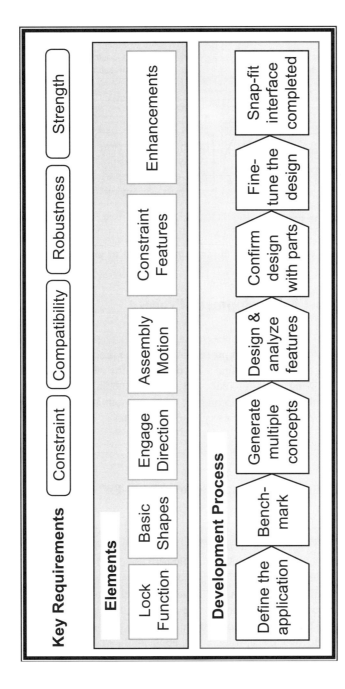

Figure 7.8 The Attachment Level™ Construct for snap-fits

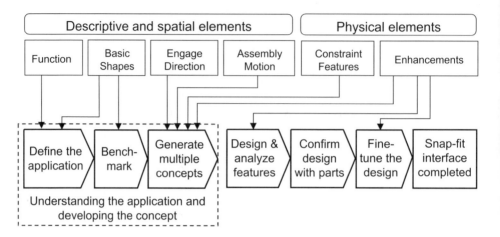

Figure 7.9 Where the elements fit into the snap-fit development process

7.2 The Snap-Fit Development Process

This section describes the development process in detail. The process is presented as a series of steps. For each step, the reader will find a discussion and the general rules associated with that step. A table showing cross-references to related material in this book or to other publications is included and, where appropriate, decision tools in the form of checklists or reference tables are provided to aid in carrying out the step.

7.2.1 Is the Application Appropriate for a Snap-Fit? (Step 0)

The snap-fit development process assumes that a snap-fit is the chosen attachment method. However, before starting to develop a snap-fit for a particular application, one must decide if the effort is likely to succeed. The checklists in Tables 7.1a–c will help the designer consider the question, "Is this application a good candidate for a snap-fit?" Use the checklists to understand the potential issues around using a snap-fit as well as a reminder of roadblocks to watch out for as you proceed. Many of the items in the list will not stop snap-fit development, but they may make it more difficult or more time-consuming. "No" answers do not necessarily rule out a snap-fit, but an application with many "yes" answers is probably a more reasonable candidate for a snap-fit. Do not let the list scare you. These are the kinds of issues that occur in many product development projects and it is a good idea to consider them at the start. Especially because, for many designers, developing a snap-fit application may be a new experience. One possible use of this list is to use it to categorize applications as low, medium or high risk. Resources for developing snap-fits can then be allocated to the applications based on risk and projected savings.

7.2 The Snap-Fit Development Process

Table 7.1a Is a Snap-Fit Attachment Appropriate? Application Considerations:

Application	Response*		Why
Do you have design responsibility for both the mating part and base part?	Yes	No	It is much easier if you "own" both parts.
Does your organization have design responsibility for both parts?	Yes	No	Communication is important.
Are manufacturing volumes high?	Yes	No	Must recover higher initial costs.
Does a validation procedure exist for the application and will it test the snap-fit?	Yes	No	End-use testing is important.
Are performance requirements available for the application?	Yes	No	Snap-fit must meet them too.
Is the application spring-loaded? Can it fly apart during assembly or service?	Yes	No	May cause injury, a "booby-trap".
Is sealing required in the application? Will gaskets be used?	Yes	No	Sealing may require clamp load.
Is clamp load required in the application?	Yes	No	Plastic snap-fits can't give clamp load.
Will high or sustained forces be applied to the attachment?	Yes	No	Increases possibility of plastic creep.
Will the application experience shock or impact loading?	Yes	No	Careful analysis and strong locks needed.
Is the application subject to mass loading only?	Yes	No	Preferred to functional or structural loads.
Is the application subject to a high frequency of service?	Yes	No	Damage or fatigue of locks is possible.
If service is required, is disassembly obvious or is instructional information available?	Yes	No	Reduce chances of damage.
Is the application used in a high temperature environment?	Yes	No	Short-term plastic performance changes and long-term degradation.
Is the application used in an extreme low temperature environment?	Yes	No	Causes brittle behavior in plastics.
Do federal safety, health or other standards regulate the application?	Yes	No	If it is, thorough documentation required.

* The response indicated in dark font is generally more favorable to use of a snap-fit.

If snap-fits are a new experience for the organization, it is critical that product engineering management understand a few basic things about snap-fits. First, the time and effort required to develop a reliable and cost effective snap-fit attachment will most likely exceed the time spent on a more traditional (loose fastener) attachment for the same application. Second, benefits that far exceed the initial engineering costs are realized when

Table 7.1b Is a Snap-Fit Attachment Appropriate? Component and Material Considerations:

Components/Materials	Response*		Why
Is the mating part high mass?	Yes	No	Stronger locks required.
Is there adequate space on the parts for snap-fit features?	Yes	No	Space for lock deflection and protrusions.
Is one or both of the parts to be made of plastic?	Yes	No	Easier to do a snap-fit in plastic.
Is the mating part a: Trim, Bezel, Panel, Control module Cover, Switch, Access door	Yes	No	These applications are usually easy.
Is either of the parts expensive?	Yes	No	Consider a back-up attachment.
Do the joined materials differ significantly in rate of thermal expansion?	Yes	No	Care needed in developing constraint.
Are the parts made of "engineering" polymers?	Yes	No	More predictable and higher performance.
Is the application exposed to ultra-violet light?	Yes	No	Performance degradation is possible.
Is the plastic exposed to chemicals in the environment?	Yes	No	Performance degradation is possible.
Is high dimensional variation likely?	Yes	No	Care needed in developing constraint.
Are you a polymers expert or do you have access to an expert?	Yes	No	Materials data interpretation.

*The response indicated in dark font is generally more favorable to use of a snap-fit.

that design is assembled thousands of times by an operator using no tools and no threaded fasteners. Third, evaluating the loose vs. snap-fit attachment in terms of the *total assembled cost* is important because the piece-cost alone of a part with snap-fits will be higher than that of a part without snap-fits.

It should be mentioned here that snap-fits are not limited to only plastic-to-plastic applications. They are also appropriate for many plastic-to-metal applications as well as some metal-to-metal applications. Details of material properties are, of course, different for metal parts, but all the rules of snap-fit performance still apply.

Once it is determined that an application is a reasonable candidate for a snap-fit attachment, the development process can begin. Note, however, that even if the designer ultimately determines that a snap-fit locking feature can not be used in the application, their time has not been wasted. The thinking process a designer goes through to create a snap-fit will result in a better attachment regardless of the final locking method. Selecting an alternative fastening method will be discussed at the appropriate step in the process.

Table 7.1c Is a Snap-Fit Attachment Appropriate? Information and Organizational Considerations:

Information/Data	Response*	Why
Will accurate load information be available for analysis?	Yes No	For critical applications, a necessity.
Is accurate material property data available for both of the parts to be joined?	Yes No	Needed for accurate analysis.
Will accurate dimensional data be available?	Yes No	For determining position and compliance.
Is part/base packaging known or predictable?	Yes No	Access for assembly motions & service.
Do you know the possibility of misuse or unexpected loads on the attachment?	Yes No	For complete analysis of reliability.
Organizational	*Response**	*Why*
Is the application a new design rather than a carry-over?	Yes No	Sometimes it is easier to start fresh.
Is there enough lead-time to accommodate possible longer design time?	Yes No	Generally a longer development time.
Does the organization understand the trade-off between a piece-cost penalty and assembly savings?	Yes No	Support for the effort.
Does the part supplier have experience with molding snap-fit applications?	Yes No	Better understanding of manufacturing requirements and issues.
Does the purchasing/bidding process allow the final supplier to be the prototype supplier?	Yes No	They will learn from prototype development.
Does the purchasing/bidding process allow the supplier to participate in design meetings?	Yes No	Can give advice during development.

* The response indicated in dark font is generally more favorable to use of a snap-fit.

Applicability of attachment level thinking to other attachment methods is briefly discussed in Chapter 2.

Because the designer's choice is frequently between use of a threaded fastener or a snap-fit, a short list of typical advantages and disadvantages of each method is appropriate. Because we are discussing plastic, these are written in terms of attaching plastic parts to each other. Some of the issues would be different for metal to metal attachments.

Advantages of threaded fasteners include:

- Robust to dimensional variation
- Allow for adjustment after assembly
- Fastener strength is independent of joined material
- Part interface is simple and initial design cost is usually lower
- Part processing is easier
- Supports low volume productions
- Part piece cost is lower
- Disassembly for service is obvious

Disadvantages of threaded fasteners include:

- Clamp load may crack plastic
- Additional parts in the product and in inventory
- Each fastening site may require as many as three additional fasteners (screw, washer, nut)
- Longer assembly time
- Capital costs for assembly tools
- Visible fastener may be undesirable
- Fasteners may strip during assembly for a hidden failure

Advantages of snap-fits include:

- Fewer parts in product and in inventory
- Lower assembly time
- No visible fasteners, clean appearance
- Can be made non-releasing and permanent
- Can give feedback of good assembly
- No investment for power tools

Disadvantages of snap-fits include:

- Parts are more complex and piece cost is higher
- Development costs are higher
- Close control of dimensions is required
- No adjustment after assembly
- Fastener strength is limited by parent material strength
- Hidden fasteners may be hard to service

Once the decision to proceed with a snap-fit is made, the development process can begin.

7.2.2 Define the Application (Step 1)

To begin the development process, the application is defined using the descriptive elements *function* and *basic shape*. Function, summarized in Table 7.2 describes the nature of the locking requirements for the attachment. Basic shapes, Table 7.3, are generic descriptions of the part's geometry.

7.2 The Snap-Fit Development Process

Table 7.2 Locking Function in the Application

Action	Movable **or** Fixed	Free movement or controlled / No movement once latched
Purpose	Temporary **or** Final	Until final attachment is made / Snap-fit is the final attachment
Retention	Permanent **or** Non-permanent	Not intended for release / May be released
Release	Releasing **or** Non-releasing	Releases with applied force on the mating part / Lock is manually deflected for release

Defining the application using these attachment level terms will help when design rules are applied later in the process. Their immediate value, however, is in helping designers structure a search for ideas as they conduct technical benchmarking in the next step.

In addition to the attachment level elements described above, each application will have specific performance requirements and in-service conditions which must be defined. Keep in mind that some of these need not be known at this stage of the process, but they will be needed eventually. In general, the sooner this information is collected, the better. Application-specific requirements and conditions can include:

- Material properties.
- Manufacturing limitations and capabilities.
- Load history for the application.
- Thermal history for the application.
- Alignment and appearance requirements.
- Environmental conditions such as chemical and ultra-violet exposure.

At this time, the designer should begin hand-drawn sketches of the application in terms of its basic shapes. These "concept sketches" will be used to capture ideas and alternatives throughout the snap-fit concept development process. The designer should also begin thinking about how a crude model of the application can be constructed.

Table 7.3 Possible Basic Shape Combinations in the Application

	Solid	Panel	Enclosure	Surface	Opening	Cavity
Mating part	Common	Common	Common	Rare	Rare	Low
Base part	Common	Rare	Rare	Common	Common	Common

Table 7.4 Cross-references for Step 1, Define the Application

Function	Chapter 2, Section 2.3.1
Basic Shapes	Chapter 2, Section 2.3.2

Table 7.4 lists cross-references for decisions made during this step of the process. Blank spaces are also provided so the reader can add additional references if desired. Throughout this chapter, cross-reference tables are provided at the ends of most of the sections describing steps in the development process.

7.2.3 Benchmark (Step 2)

The term benchmarking has many meanings. In the snap-fit development process, it is not marketing, customer or product feature benchmarking. It is *technical* benchmarking and it means the careful study of other applications for understanding, learning and ideas. It is not simply reverse engineering in order to copy other designs or ideas. Simply copying without understanding can lead to problems, both technical and legal. Benchmarking is a continuous process of learning and changing [6]. When one truly understands the attachment method on another product, the tendency is to change and improve on it for one's own application. The idea of benchmarking is to stimulate creativity and ideas by becoming familiar with some of the available design options. The products studied should include your own company's as well as your competitors' products. It is also important to study products that are unrelated to your own product or to the application under development. This is one reason why the concept of basic shapes is so important. By describing an application in terms of its basic shapes, the designer is free to seek ideas in any products having similar basic shapes.

This is one of the important and basic principles of the process: Benchmark by studying other products that have the same basic shapes as your application. The designer is not limited to studying only applications that are similar to the one being developed. The field of study is opened up to any product and parts that have the same basic shape combination. Creative ideas become available everywhere.

The worksheet in Table 7.5 can be used as a reminder of snap-fit features and attributes to look for when benchmarking.

One of the simplest of applications is a rectangular panel to an opening. If the reader was to study a number of panel-opening applications, they would discover a great variety of design interpretations. Obviously, some will be fundamentally better than others and some will be better for a given application. A designer can choose to invent a new panel to

Table 7.5 Benchmarking Checklist

For development project:	Comments:
Application(s) studied:	
Is the application properly constrained? 　How is it constrained? 　Are the features effective?	
Any evidence of damage? 　Stress marks? 　Damage to edges or corners?	
How does it feel? 　Assemble it, is it easy? 　Shake it, any noises or movement? 　Drop it to the floor (maybe)	
Take it apart. 　Could a customer take it apart without damage? 　Are tools required for disassembly?	
Look for all enhancements, especially the required ones. Are they all there? Are there any shortcomings that need enhancements?	
Is it easy to make? 　No die lock 　Simple features	
How would you rate the application if you had to assemble it 8 hours a day?	Poor　Fair　Good　Excellent
How would you rate the application if you were a service technician?	Poor　Fair　Good　Excellent
How would you rate the application if you had to manufacture it?	Poor　Fair　Good　Excellent
How would you rate the application if you were a customer?	Poor　Fair　Good　Excellent

opening for their application, or they can study existing applications and select from the best ideas found as well as the best ideas they generate for themselves. This example also points out another extremely powerful aspect of defining an application in terms of basic shapes. That is, "Why re-invent a new concept?" Once an excellent panel-to-opening design concept has been created, simply adapt it to the application at hand. A designer can establish a library of good concepts for a variety of basic shape combinations and draw upon that as a reference for new design.

Benchmarking is easy to ignore, but it is extremely important in driving creativity. As familiarity with snap-fits increases, attachment level benchmarking will occur naturally when products are studied.

7.2.3.1 Rules for Benchmarking

- Benchmark on basic shapes and do not limit the search to just one type of product or application. Many plastic products are available for study including toys, electronics,

small appliances, etc. Most of the time, ideas drawn from several applications will influence the final design.
- Make it a point to look for enhancements. Enhancements are often added to the product after the fact because a problem was discovered. When one can recognize enhancement features and understand what they do, they can either be included in the attachment design from the start or the condition that made the enhancement necessary can be avoided.
- Study the constraint features and how constraint and compliance issues are resolved. Understand why the locks and locators were selected and arranged as they are. Study how they behave and interact as the parts are brought together through the required assembly motion.
- Assemble and disassemble the parts. How do they feel? Shake the parts. Do they squeak and rattle? Are the parts stiff enough? Look at the distribution of constraint features on the parts. Are there enough to compensate for part flexibility? Flexibility is of particular concern with large panels as they are usually weak in bending.
- Study the parts for witness marks indicating over-stress or assembly problems. A lighter color or whitened area at the base of a lock or locator indicates damage. Broken or damaged edges or corners on parts indicate interference and difficult assembly. These are clues to strength requirements as well as enhancements that may be needed.
- Cross-references for the benchmarking step are shown in Table 7.6.

7.2.4 Generate Multiple Attachment Concepts (Step 3)

Figure 7.9 showed how the elements of a snap-fit map to the development process. In that diagram, note that four of the six elements are brought together in the third step. This step is the most critical of the process because it is where most of the important decisions about the attachment are made and because it helps to ensure creativity during snap-fit development. By identifying combinations of allowable engage directions and assembly motions, the designer can create *several fundamentally different attachment concepts* rather than mentally locking themselves into only one idea or just variations on one basic theme. Thomas Edison said, "To have a good idea, have lots of them". Developing concept alternatives is an

Table 7.6 Cross-references for Step 2, Technical Benchmarking

Constraint Features	Chapter 3
Enhancement summary table	Chapter 4, Table 4.2
Required enhancements	Chapter 4, Table 4.3
Enhancements and the development process	Chapter 4, Table 4.4

important enabler for creativity [7]. Constraint features and some enhancements are then added to each alternative, the concepts are evaluated and one is selected for analysis and design.

Step 3 need not be long or difficult and it can be conducted as a personal or as a group brainstorming session. Knowledge gained during careful (attachment level) definition of the application (Step 1) and benchmarking (Step 2) is now applied as attachment concepts are generated. Step 3 consists of the five sub-steps shown in Fig. 7.10.

Important: This step is highly recommended as an activity during design for assembly workshops. The mechanics of using assembly motion alternatives to force generation of fundamentally different attachment solutions is a critical part of the development process.

7.2.4.1 Select Allowable Engage Directions (Step 3.1)

Once the application's design constraints and shapes are defined, engage direction is generally the first decision made in the snap-fit design process. (The reader should understand that *engage direction* is not the same as *assembly motion*.) Careful selection of an engage direction is important because it is associated with a separation direction that, in turn, determines locking feature orientation. In most snap-fit applications, the separation direction is opposite the engage direction. In other words, the lock pair(s) disengage or separate in the opposite direction from which they engage. The basic rules for selecting allowable engage directions are:

- Engage directions must be compatible with the basic shapes.
- They must be compatible with access for assembly, service and usage. Also consider if any other parts added later would interfere with service.
- They must be ergonomically friendly if the parts will be assembled by human operators.

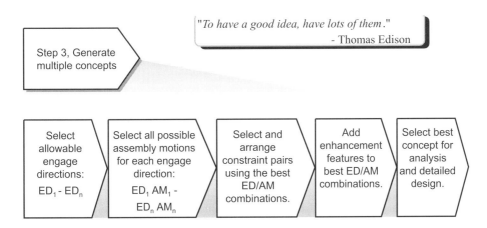

Figure 7.10 Details of the generate multiple concepts step

- If intended for automatic assembly, consider the impact of access and motion complexity on capital equipment costs.
- There should be no significant forces acting in the separation direction.

The last point deserves additional comment. Any forces acting in the separation direction will be acting directly against the lock features, trying to pull the components apart. Because the lock features are inherently weak, this is always an undesirable situation. (However, we have seen that there are ways to make the lock feature stronger so they can carry some forces.) How does one know when the separation forces are significant? Analysis of the lock feature performance will give an indication, but it is a good design practice to simply avoid any forces in the separation direction if possible. Keep in mind also that the significance of a force may depend on more than just its magnitude, duration and frequency are also important considerations.

Forces across the snap-fit interface are one of the application specific requirements that the designer should know. Early in the development process, information on the magnitude of the forces may not be available or force information may be based on estimates with more exact data to come later. At this stage of development, it is not necessary to know the exact magnitudes of the forces. It is important to know the direction and a relative magnitude of each force. Most of the time, engage direction decisions can be made with this information.

Some applications will have only one allowable engage direction, others will have more than one. All allowable engage and separation directions should be represented by vectors on a convenient coordinate system selected by the designer. This coordinate system can now be added to the concept sketches begun in Step 1. The engage and separation vectors are also added to the sketches now. Some applications will have more than one possible set of engage and separation directions and all allowable directions should be identified. In Fig. 7.10, they are identified as ED_1, ED_2, etc.

7.2.4.2 Identify All Possible Assembly Motions (Step 3.2)

Recall that five simple assembly motions: push, slide, tip, spin and pivot are used to describe final mating part motion as the lock(s) engage. Each assembly motion will allow certain constraint feature configurations and preclude others. In Fig. 7.10, assembly motion is identified as AM_1, AM_2, etc.

Given the allowable engage directions determined in Step 3.1 and the basic shapes, the designer will find that only some of the five assembly motions are feasible. Identify all possible assembly motions for each allowable engage direction. Again, certain application conditions may render some of these combinations undesirable although they may be feasible. Eliminate those combinations from consideration. By the end of this step, one has identified some "best" combinations. The application conditions that drive assembly motion decisions include many of the same conditions that drive engage direction decisions:

- Assembly motions must be compatible with the engage direction.
- They must be compatible with the basic shapes.
- They must be compatible with access for assembly, service and usage. Also consider if any other parts added later would interfere with service.
- They must be ergonomically friendly if the parts will be assembled by human operators.

- If intended for automatic assembly, consider the impact of access and motion complexity capital equipment costs.

Figure 7.11 shows how different assembly motions will force different interface designs for the same solid to surface application. Note that, by definition, the assembly motion is the final motion made to engage the locking feature(s). For a push motion, Fig. 7.11a, all initial engagement must occur with lock features. In contrast, Fig. 7.11b shows how, for a tip assembly motion the lug(s) must be engaged first before the tip motion can begin. In Fig. 7.11c, a slide motion, the mating part must first be placed against the surface so that the lugs are aligned with the edges they will engage.

Any additional motions required to bring the mating part to the base part are not considered at this time because, although they may be affected by or related to the final assembly motion, they do not drive the constraint feature decisions.

(a) For a solid to surface attachment, a push motion forces the use of deflecting features at certain sites

(b) Another assembly motion (tip) forces the use of different features at some sites

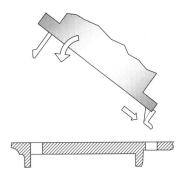

(c) A third assembly motion (slide) also forces the use of different features

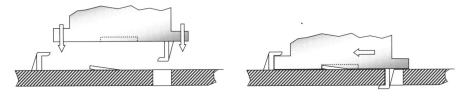

Figure 7.11 How different assembly motions force creation of fundamentally different attaching options

Rules for Selecting an Assembly Motion

- In general, the push assembly motion will likely result in a weaker attachment because more degrees of motion must be removed by the (generally weaker) lock features. The other assembly motions allow more degrees of motion to be removed by the (stronger) locators and are generally preferred.
- The tip motion has certain advantages over some of the others. The first locator pair, once engaged, stabilizes the mating part relative to the base part for easier engagement of the remaining constraint pairs. The tip motion also minimizes potential for simultaneous engagement of constraint features.
- Disadvantages of the tip motion are that the rotational movement may require more space than is available for assembly and if excessive rotation is involved, assembly operators may be subject to cumulative trauma injury.
- In general, the tip and slide assembly motions are preferred over the push motion. However, at this stage of the process, the intention is to use assembly motion alternatives to generate ideas. All feasible assembly motions should be considered at this time.

It is desirable to have at least three ED/AM combinations at the end of this step although in some applications this will not be possible. At this time, the designer should make enough copies of the original concept sketch so that constraint pairs and eventually enhancements can be sketched onto each available ED/AM combination.

Table 7.7 shows the cross-references for the Engage Direction and Assembly Motion steps.

7.2.4.3 Engage Directions, Assembly Motions and Worker Ergonomics

The subject of assembly operator ergonomics is far beyond the scope of this book, but it deserves mention at the awareness level. Readers are encouraged to seek out appropriate application specific information to ensure their final design is "operator friendly". Information related to maximum allowable assembly forces, assembly direction, workpiece height, cycle times and operator motions is particularly applicable to snap-fit design. Simplicity of design and the complexity of decisions required during assembly can impact mental fatigue, number of mistakes and product quality.

Table 7.7 Cross-references for Step 3.1 and Step 3.2, Engage Directions and Assembly Motions

Engage Direction	Chapter 2, Section 2.3.3
Assembly Motion	Chapter 2, Section 2.3.4

Some very general rules for ergonomic design include the following: [8, 9, 10]

- Product designs having low cycle times repeated over extended periods of time should have low assembly forces.
- The operator should not be required to strike or pound (as with the palm of the hand) the mating part to cause it to snap to the base part. Any impact is undesirable.
- Designs that force the operator into an unnatural body position or force an unnatural arm, shoulder, wrist or hand position while applying assembly force should be avoided.
- Avoid designs that require continuous, rapid and repetitive application of force.
- The areas of a part where the operator must apply assembly force should distribute pressure over a sufficient area of the finger or hand. Pushing against edges, corners or points should be avoided.
- Assembly of parts while wearing gloves can have negative ergonomic effects as well as process efficiency effects. If gloves must be worn, the part design must reflect that requirement.
- Part designs that favor right-handed over left-handed workers should be avoided.

Specific work-related musculoskeletal disorders (WMSDs) [10] that can be related to snap-fit assembly include: carpal tunnel syndrome, epicondylitis (tennis elbow), neck tension syndrome, shoulder tendonitis, tendonitis, ulnar artery aneurysm, ulnar nerve entrapment and DeQuervain's syndrome.

7.2.4.4 Select and Arrange Constraint Pairs (Step 3.3)

Constraint features are lock and locator pairs that prevent relative movement between parts. Ideas gained during benchmarking will now help the designer select the best constraint features for the application. The strategy of identifying several engage direction/assembly motion combinations will now pay off. As constraint features are selected and arranged, basic shape and assembly motion interactions will force the use of different constraint feature styles for each possible ED/AM combination. This drives creativity by forcing development of fundamentally different attachment concepts rather than just variations on one theme.

To this point, the designer should have been working with *hand-drawn* concept sketches. (Leave that computer and the design programs alone!) It is now time to create a 3-D model of the application. Again, as with the concept sketch, the purpose is to invoke spatial reasoning skills and creativity. The model at this stage of the process need not (in fact, it can not be) very detailed or even accurate. The most important thing is to get something in three dimensions that can be held and manipulated in space. While making constraint features decisions, use the model(s) to visualize the interactions of the mating and base parts under the different ED/AM combinations. Also use the model to visualize how the interface will react to input forces.

Models can be very useful as a visual device when explaining or trying to sell an idea to others. The highly spatial and sometimes complex nature of snap-fits can make them difficult to explain with words alone or 2-D drawings. The designer willing to provide a model has a better chance of getting their point across.

Rapid-prototyping technology makes it tempting to produce detailed models early in the development process. With some applications, this may be desirable and helpful. In many other cases, however, the effort and expense of producing these models is better left until later in the process. The creative advantages of creating a hand-made model are also lost when only machine-made models are used. Even when rapid-prototype models are indicated at this stage of the process, some crude hand-made models should also be built. Some possible models and modeling materials are listed here. They are simple and may appear trivial, but they can be a powerful tool for generating creativity:

- Cardboard or heavy card stock can be cut, glued or taped.
- Styrofoam can be carved.
- Wood can be cut and shaped.
- Craft material can be formed and fired to harden.
- Plaster of paris can be molded and cut, filed and sanded to shape.
- Scrap parts having a similar shape can be cut and shaped.
- A closed box, book or coffee cup may serve to model a solid.
- A table top may represent a surface.
- A piece of card stock can be a panel.
- An open box can be a cavity or an enclosure.
- Constraint features can be cut from card stock and glued onto the models.

a. Adding Constraint Pairs

Consider the first ED/AM combination and begin selecting and arranging constraint pairs. Note that reference to a singular constraint pair may actually include multiple constraint pairs when they are acting in parallel, Fig. 7.12. The judgement of whether to consider a constraint pair as one pair or as multiple pairs is up to the designer. At a pure qualitative level, there is no clear difference between identical constraint pairs acting in parallel and in the same sense. More quantitative math-based evaluation of the relations involved is necessary and is beyond the scope of this book.

The first constraint pair added to the application may not necessarily be the first constraint pair engaged during assembly. It should be the most constraining locator pair possible and must be compatible with the selected engage direction and assembly motion (ED/AM). This constraint pair should become the datum for locating all remaining constraint features. Add it to the concept sketch and note all the constraint vectors associated with this pair.

The second constraint pair added will also be a locator pair and it must be compatible with the first constraint pair as well as the ED/AM. The second pair should be less constraining than the first and none of its constraint vectors should be coincident with those of the first pair.

The third constraint pair added may be a locator pair or a lock pair depending on the application. It must be compatible with the first two constraint pairs and the ED/AM. If it is a lock pair, then the step is completed.

If necessary, add a fourth constraint pair. This will be a lock pair that constrains only in the removal direction. It too must be compatible with the ED/AM and all other constraint pairs.

(a) Locators

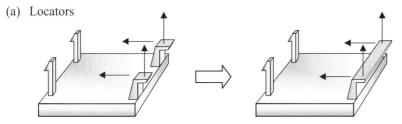

The lines of action and the net effect on constraint are the same in both cases

(b) Locks

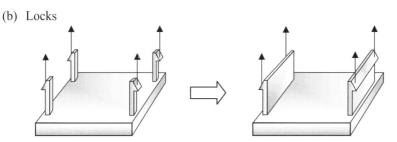

The lines of action and the net effect on constraint are the same in both cases

Figure 7.12 **Multiple vs. single constraint features**

Repeat the process for all remaining ED/AM combinations.

Some designers have an intuitive feeling for constraint and will quickly understand the process. For others, some practice is required in order to reach that understanding of constraint. A matrix of constraint pairs and constraint directions as shown in Table 7.8 can serve as a valuable learning tool. The matrix is also useful when explaining the rational behind a snap-fit design to someone else. With constraint pairs listed vertically and the 12 DOM listed across the top, the matrix can serve as a checklist for recording DOM as they accumulate with the addition of constraint pairs and for verifying that neither over nor under-constraint has occurred. The constraint matrix is discussed in more detail in Chapter 5.

Decisions about placing a given constraint feature on the base part or the mating part are driven by many considerations. Manufacturing considerations, material strength, basic shape and assembly motion are the most common. Another may be the relative value of the parts and the chances of feature (particularly lock) breakage. Design the application so that if a feature does break, (improper disassembly, overload, etc), the part likely to break is easily replaced, inexpensive, easily serviced or repaired.

b. Rules for Selecting and Arranging Constraint Pairs

Many of these rules were introduced along with constraint features in Chapter 3. Refer back to that chapter for details.

Table 7.8 Constraint Worksheet Original

Interface Requirements		Degrees of Motion											
		Translation						Rotation					
		+x	−x	+y	−y	+z	−z	+x	−x	+y	−y	+z	−z
Identify desirable axes for mating part stability, translation and rotation								(y axis) (z axis)	(y axis) (z axis)	(x axis)	(x axis)	(x axis)	(y axis)
Uni-directional effects	F_M resulting from accelerations and part mass												
	F_F resulting from functional loads												
	F_N resulting from non-standard loads												
	Engage Direction (ED) & Assembly Force (F_A)												
	Separation Force (F_S)												
Bi-directional effects	Thermal expansion / contraction												
	Alignment directions												
	Part compliance												
Constraint pairs													
Resolve bi-directional requirements	Part-to-part alignment												
	Compliance sites												
	Fine-tuning sites												

X = difficult or not available
A = available
✓ = necessary or required

- Lock pairs should constrain in as few degrees of motion (DOM) as possible and locator pairs in as many DOM as possible. Ideally, the lock pair should constrain only in the one DOM associated with part separation.
- Generally, the more degrees of motion removed with locators, the stronger the attachment.
- A tip or slide assembly motion is preferred over the push motion because more degrees of motion are removed with locators and because of ease of assembly.
- The application should not be over-constrained due to redundant constraint pairs.
- Over-constraint due to opposing constraint pairs is undesirable but sometimes necessary.
- Where constraint pairs oppose each other (two constraint pairs with collinear strength vectors of opposite sense), placing the pairs as close to each other as possible will minimize tolerance effects and the potential for opposing internal forces within the constraint system.
- Where constraint pairs create a couple, (parallel strength vectors of opposite sense) they should be placed as far apart as possible for mechanical advantage and reduced sensitivity to tolerances.
- Where constraint pairs have parallel strength vectors (of either the same or of opposite sense) they should be placed as far apart as possible for maximum mechanical advantage and reduced sensitivity to dimensional variation.
- Applications should not be under-constrained. Under-constraint occurs when no constraint pairs provide strength in one or more translational degrees of motion or when a constraint couple is ineffective in removing rotational constraint.
- In a fixed application, the mating part must be constrained to the base part in exactly 12 DOM. An exception is certain functional attachments where free movement is allowed in some degrees of motion.
- Locking features should not carry high forces or sustained forces, particularly in the separation direction or in bending.
- Compliance should generally occur within a constraint pair rather than between pairs.
- The lock retention face can be used to take up some tolerance. A slight angle on the retention face of a (90° non-releasing) hook will absorb tolerance without affecting retention. A contoured face can ensure maximum retention angle at any level of hook deflection.
- Select and orient constraint pairs whenever possible to avoid a die-lock condition.

After adding constraint features to the concept sketches, the designer may also wish to build them into the 3-D models. Rapid prototype models may again be considered. Including the locator features on models is useful because their presence allows one to evaluate ease of assembly and some aspects of constraint. In many cases however, there is little or no value in including the (flexible) locking features on the models. Sometimes materials used in rapid-prototyping are brittle and the flexible features are soon broken off. It is possible to make locking features out of flexible plastics and attach them to the models with screws or adhesives. This can be useful in some cases although these locks will not represent realistic lock feature performance.

Table 7.9 is the cross-reference for selecting constraint features.

Table 7.9 Cross-references for Step 3.3, Select Constraint Pairs

Constraint introduction	Chapter 2, Section 2.2.2
Constraint concepts	Chapter 5, Section 5.1.2
Locator feature styles	Chapter 3, Section 3.2.1
Locator pairs	Chapter 3, Section 3.2.2
Lock feature styles	Chapter 3, Section 3.3.1
Design rules	Chapter 3, Section 3.4.2

7.2.4.5 Add Some Enhancement Features (Step 3.4)

For each concept alternative, decide which enhancements are needed. Enhancements are either physical features or attributes of constraint features or of the parts themselves. At this step in the development process, one can often predict the need for some enhancements depending on the nature of the application. Guides, pilots, visual, assists and guards are enhancements that can usually be added to the attachment concept now, if the application requires them. The remaining enhancements are normally added later when a detailed design is established.

Table 7.10 is the cross-reference for adding some enhancement features.

7.2.4.6 Select the Best Concept for Feature Analysis and Detailed Design (Step 3.5)

To this point, the development process has been both a structured and a creative process. The result is several fundamentally different and technically sound snap-fit attachment concepts. Each concept has constraint features arranged to provide proper mating part to base part constraint. Some enhancement features have also been added to each concept.

At this point, each concept should be reviewed by appropriate stakeholders. Likely stakeholders include the product engineer(s) and designer(s) for both the mating and base parts; cost analysts, purchasing agents, materials experts, part manufacturers and manufacturing, assembly and process engineers. The best concept is selected to be carried forward into design and recommendations for improvements may also be made. The other concepts, if judged feasible, can be ranked in order of preference and kept available should the selected design become unacceptable as the program proceeds. The models and sketches created to this point should be available for this concept review and can be valuable tools for

Table 7.10 Cross-references for Step 3.4, Add Some Enhancement Features

Enhancements	Chapter 4
Enhancements summary	Chapter 4, Table 4.2
Enhancement requirements	Chapter 4, Table 4.3
Enhancements and the development process	Chapter 4, Table 4.4
Design rules	Chapter 4, Section 4.6.2

explaining the details of each design. Table 7.11 is a worksheet for comparing alternatives when selecting the best concept.

Whenever possible, plan for the prototype supplier to also be the production supplier. Plastic part tooling and processing requires a thorough understanding of the application. Supplier input during the initial design can help ensure a functional design that can be reliably produced. Single sourcing of parts may be desirable for the same reasons. These kinds of decisions are often made based on purchasing and organizational procedures, but the designer may be able to present a solid business case for a single knowledgeable supplier.

This concludes the concept development phase of the process. We exit Step 3 with a fundamentally sound attachment concept ready for snap-fit feature analysis and detailed design. We now move into the more familiar and traditional area of feature analysis and detailed design.

7.2.5 Feature Analysis and Design (Step 4)

To this point, we have been working with concepts and ideas, not dimensions or details. Step 4 is the detailed design step. The objective is to evaluate feature performance and determine dimensions for:

- Acceptable installation and removal effort.
- Retention and load carrying strength.
- Acceptable stress and strain levels during assembly and release removal deflection.
- Acceptable stress and strain levels under applied loads.
- Squeak and rattle resistance.

Table 7.11 Worksheet for Step 3.5, Select the Best Concept

	Attachment alternative		
	#1	#2	#3
Constraint execution			
Efficient use of features			
Meets minimum requirements for enhancements			
Ease of assembly			
Estimated piece cost			
Supports the business case			
Ease of manufacturing			
Meets business ergonomic requirements			

Note that analysis of any kind is of limited value unless the snap-fit interface is properly constrained.

In some cases feature sizing is carried out based on experience and analysis is used only if testing indicates a need for it. In other cases analytical methods are applied immediately to evaluate feature performance and determine feature dimensions. Step 4 is the traditional (feature level) snap-fit technology.

Chapter 6 contains some of the more common rules of thumb for sizing cantilever style locks. It also provides some of the more common calculations for evaluating feature performance. However, because analysis of performance has represented snap-fit technology for so long, there are already many good sources of feature analysis information in published design guides, technical reports and commercial software tools. Sources for analysis information are listed in the references at the end of Chapter 6.

Analysis may indicate that the selected features can be designed to meet all application requirements. Analysis may also indicate the need for additional constraint features for increased strength or retainer enhancement features to improve retention beyond the inherent strength of the lock pair(s) in question. The results may also indicate that the selected concept cannot be designed to meet the required objectives and that one of the alternative concepts should be tried.

Recall that the purpose of the first three steps of the process is to create a fundamentally sound attachment concept. A concept can be sound and yet fail to meet one or more of the objectives because of a combination of material performance limitations and conflicting performance requirements. For example, a high retention strength requirement can be in conflict with a low insertion force requirement or the material properties will not support the required assembly performance. Why then should we waste our time creating a fundamentally sound attachment if it might not work? Because the fundamentally sound concept has the best chances of working.

Additional enhancements should be added at this time if indicated. In all applications, the manufacturing enhancements (process-friendly and fine-tuning) should be included in the final design.

7.2.5.1 Lock Alternatives

The purpose of a snap-fit is to use integral lock features. But, the feature analysis may indicate that any integral lock cannot be made to work in the application and reality is simply that snap-fits will not work everywhere. All is not lost, however, because other locking methods are often available for use in place of the integral lock. If the snap-fit design process has been followed, a properly constrained attachment with a number of locating features now exists. The final decision about lock dimensions is not made until this point. When we recall that lock features are the last constraint features added to the design and the last features engaged during assembly, we see that, for most applications, replacing an integral lock with an alternative locking feature can be relatively simple.

Details of the fastening methods suggested as alternatives to integral locks are beyond the scope of this book. The purpose here is simply to introduce the idea of alternatives to the integral lock and give the reader a starting point for further investigation.

Some of the lock alternatives described here lend themselves to automatic assembly to the parent component. In that case, they can be installed before final assembly and, as far as assembly operators are concerned, they are integral locks and the assembly labor savings apply just as with integral locks. Even screws can sometimes be pre-assembled to plastic parts so the operator does not need to handle loose fasteners. The mechanism used to capture the screw in the part can be a snap-fit feature.

Recall also that provisions for back-up fastener(s) can be designed into an application as a performance enhancement.

Any separate fastening method will add some cost to the attachment, but by following the process and designing for proper constraint using locators, the number of loose fasteners is minimized as is the associated cost impact. Three common lock alternatives, screws, push-in fasteners and spring clips, are discussed here. Other attachment methods like hook-loop fasteners and double-back tape may also be considered as alternatives to an integral lock feature.

a. Screws

In place of an integral lock, a loose threaded fastener can be used. If locators are identified and the attachment is properly constrained, usage of loose fasteners will be minimized and the design will be optimized for design for assembly.

If a screw is to be used, the first decision to be made is about the internal thread. Will it be made directly in the plastic material or will it also be a separate part? If the threads will be made directly in the plastic, the method of making the threads must be considered.

Do not use sheet metal or machine thread fasteners for tightening directly into plastic. Use fasteners that are designed specifically for tightening into plastic. These screws generally use various combinations of special thread form, thread pitch or cross-section shape to reduce the stresses produced in the plastic as material is displaced to form the threads. Other styles have cutting flutes that cut the plastic away to form the thread. Selection of a thread forming or a thread cutting screw should be based on the application's tolerance for chips (created if thread cutting screws are used) and the properties (hardness/toughness are most important) of the material in which threads are to be created.

Some suppliers of fasteners for use in plastic are listed in the cross-reference, Table 7.12 at the end of this chapter, but this list is far from all-inclusive. Running screws directly into plastics also requires careful design of the area of the part intended to accept the screw. Running a screw into a thin wall is generally undesirable and a boss should be added for additional length of thread engagement. Boss design is beyond the scope of this book but boss design guidelines are available from the resin suppliers and usually from the screw manufacturers as well. A few general considerations for using screws in plastics are:

- A boss will tend to cause a sink mark on the opposite side of the wall on which it sits. Choose the depth of core pin and boss wall thickness carefully.

Table 7.12 Cross-references for Step 4, Feature Analysis and Design

Lock feature rules of thumb	Chapter 6, Section 6.3
Analytical methods	Chapter 6, Section 6.9
Designing plastic parts for assembly	Tres, Paul [11]
Plastic part design for injection molding	Malloy, Robert [4]
Screws for use in plastics	Camcar-Textron, Rockford, Illinois ITW—Shakeproof, Elgin, Illinois
Plastic push-in fasteners	ITW—Deltar, Frankfort, Illinois TRW Fastening Systems, Farmington Hills, Michigan
Metal spring clips	California Industrial Products, Livonia, Michigan Eaton Corp. Cleveland, Ohio

- A boss can create high residual stresses and/or voids at its base which will weaken the area.
- If the screw enters the free end of the boss rather than at the base, allow for a stress relief area at the end of the boss by recessing the pilot hole slightly. This is easy to do because the hex removal feature at the base of the pilot hole core pin can provide this recess.
- Screws with countersunk heads should not be used against plastic because the wedging action of the screw head will tend to crack the material.
- Screws coated with oil or having an oil-based finish should not be used in plastic because some plastics degrade in the presence of oil.
- Distribute the pressure under the head of the screw over a wide area with a captive washer on the screw.
- If high speed tightening of the screw is the assembly method, a captive washer is recommended for screws tightened against plastic. Heat build-up due to friction under the head of the screw may melt the plastic if a washer is not present.
- High speed tightening of the screw may also cause heat build-up in the plastic material to the point that the properties of the plastic in the area of the threads will degrade resulting in very weak threads. Limiting the tightening speed may be necessary.
- Screws will develop clamp load. Plastics tend to creep and a high clamp load may result in long-term cracking of the plastic under the fastener and/or in the boss.
- Over-tightening and stripping of the plastic threads is possible. This is a hidden failure and it may leave the assembly plant undetected.
- If screws are to be removed and reassembled into the plastic multiple times for repair, cumulative damage to the plastic threads is a possibility.
- If the application does not permit perfect alignment of the screw to the pilot hole for assembly, then cracking of the boss during rundown is likely.
- The driving impression style on the screw head must be selected for screw stability during rundown. The driver bit must be selected for compatibility with the driving impression.

When the feasibility of threading directly into plastic is questionable, another method of attaching to plastics with threaded fasteners involves use of molded in or pressed in metal inserts having machine threads. These inserts add cost to the process and to the parts but provide higher thread strength. They also provide a solution where the extra thickness of a boss is not possible and the screw must run into a wall.

Screws can also be used with separate internal threads like machine thread nuts and single or multiple thread impression nuts and clips. As with inserts, this eliminates the considerations associated with threading directly into plastic bosses, but the clamp load considerations still apply.

Sometimes loose fasteners can be captured in the plastic parts prior to final assembly. This eliminates operator handling and saves time. Some methods for capturing screws in parts are shown in Fig. 7.13.

References [4] and [12] provide good discussions of many of the issues associated with loose fasteners in plastics.

248   The Snap-Fit Development Process   [Refs. on p. 253]

Figure 7.13   Threaded fasteners captured in plastic parts

b. Push-In Fasteners

We do not tend to think of push-in fasteners as snap-fits, but in many ways they are. Push-in fasteners are usually spring steel or plastic, Fig. 7.14, and they involve integral feature deflection and return for interference, just like a snap-fit lock. Like snap-fits, push-in fasteners do not generate significant clamp load. The common one-piece style push-in style fasteners do not decouple assembly and retention but the two-piece fasteners do. (See Chapter 5.)

Sometimes, push-in fasteners can be installed automatically in the part before reaching final assembly. In this case, as far as the assembly operator is concerned, they are integral snap-fits. As a designer, keep in mind that all the rules for good snap-fit design will apply to these applications. Other push-in fasteners are installed by the operator and will add cost as a separate part in the assembly process. Ergonomic limits on push-in forces apply to these fasteners when they are hand-installed. Removal for service can also be an issue. The

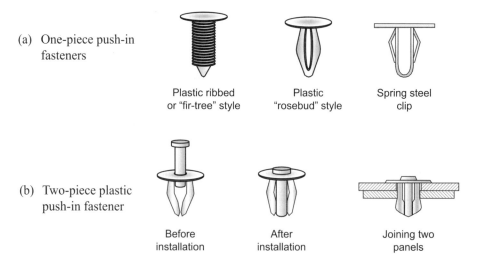

Figure 7.14   Examples of push-in fastener alternatives to integral lock features

popular ribbed plastic fasteners sometimes are difficult to remove and may be damaged during removal. Many one-piece styles will also damage the mating parts during removal unless the parts are designed to be quite strong in the area of the push-in fastener. The two-piece push-in styles can be very easy to remove when a provision is designed into them to assist in removal.

For some applications that use screws, simply replacing screws with push-in fasteners should be considered as an intermediate step in conversion to integral snap-fits. The holes already provided for screws can be used as attachment sites for push-in fasteners.

c. Metal Spring Clips

These fasteners are designed to grip adjoining features on the parts to hold them together. When installed automatically, these fasteners enter final assembly already attached to a part. As with pre-installed push-in fasteners, the application is a snap-fit as far as the operator is concerned and, once again, all the rules of snap-fit design apply. These fasteners are frequently designed with sharp barbs intended to dig into and hold the parts together. The barbs can be effective, but they will cut grooves into the plastic when removed. This makes them less effective in retention when reassembled. Other clips grip with a spring-like action and are friendlier to disassembly and reassembly.

Table 7.12 is the cross-reference for Step 4, Feature Analysis and Design.

7.2.6 Confirm the Design with Parts (Step 5)

In this step, the first parts are produced and evaluated. Results of this initial evaluation will likely indicate the need for modifications to the design. Specific application performance requirements will determine how these parts are evaluated, but it is essential that an evaluation include:

- Evaluation of ease of assembly, including access, motions, assembly force and operator feedback.
- Evaluation of user-feel if the customer will be operating the snap-fit frequently.
- Verification that the design is process-friendly.
- Verification of serviceability, if required.
- Verification of proper constraint by the locators and locks.
- Verification of strength and resistance to squeak and rattle.

The results of the evaluation and testing will indicate whether or not changes must be made to the design. The thorough understanding of the application that results from following the snap-fit development process should make it easy to identify the changes needed. Chapter 8 also describes a diagnostic process for investigating problems and recommending changes to snap-fit applications.

7.2.7 Fine-Tune the Design (Step 6)

In this step, changes indicated by evaluating first parts are made. Any necessary mold changes will be much easier if fine-tuning enhancements (Chapter 4) have been included in the design. Other enhancements that now appear to be necessary can be added. Of course, several cycles of design confirmation with parts and fine-tuning may occur before the part attachments are acceptable. One objective of the snap-fit development process is to reduce the design iterations required to get good parts.

7.2.8 Snap-Fit Application Completed (Step 7)

The attachment level development process is now completed. As a final review, the evaluation worksheet in Table 7.13 can be used to confirm that all the important aspects of attachment level snap-fit development have been considered.

7.3 Summary

This chapter explained the snap-fit development process in detail. It also provides cross-references to other areas of the book and other sources of information related to the process. The snap-fit development process must be conceptual and creative before it is analytic. By design, the attachment level construct is rule-based to support learning and practical application of profound knowledge to snap-fits. The step-by-step development process described here leads the designer to apply those rules. A snap-fit designer will experience improved creativity and spatial reasoning when developing attachment concepts. As the process is followed, it becomes second nature as users grow in their understanding of snap-fits.

7.3.1 Important Points in Chapter 7

- The snap-fit development process is highly compatible with and supports design for assembly (DFA) principles. Specific aspects of this process should be included in the design for assembly thought process and in DFA workshops. The concepts of generic basic shapes and assembly motions as well as the use of assembly motion to drive interface design alternatives are particularly important and applicable to design for assembly [13].

7.3 Summary

Table 7.13 Final Snap-Fit Evaluation

Basic shapes	Mating part is	Solid Panel Enclosure Cavity Opening
Function	Base part is Action is Attachment type is Retention is Lock type is	Solid Enclosure Cavity Opening Fixed Moveable (free/controlled) Temporary Final Permanent Non-permanent Releasing Non-releasing
Assembly motion	Preferred	Push **Slide** Tip Twist Pivot*
Strength/forces	Requirements/directions identified?	Yes* No
Alignment	Requirements/directions identified?	**Yes** No
Packaging	Operator access Clearance for part movements	**Easy** Difficult **Yes** No
Material	Properties/families identified	**Yes** No
Geometry	Dimensions and tolerances identified	**Yes** No
Separation direction	Any significant forces?	Yes **No**
If significant forces in separation direction	Retainer enhancements added? Trap locks used? Decoupling considered?	**Yes** No **Yes** No **Yes** No
Locators	Used as guides? Used as pilots? Usage is maximized? Have clearance enhancements? Well distributed on panels to prevent squeak and rattle? Critical pair selected as datum for others? Simultaneous engagement of multiple locator features?	**Yes** No **Yes** No **Yes** No **Yes** No **Yes** No **Yes** No Yes **No**
Locks	Usage is minimized? Carrying high or sustained forces? Well distributed on panels to prevent squeak and rattle? Simultaneous engagement of multiple no features? Simple shapes? Engage locators in the lock pair? Only constrain in the separation direction?	**Yes** No Yes **No** **Yes** No Yes **No** **Yes** No **Yes** No **Yes** No
Benchmarking	Conducted on	Basic shapes Similar applications **Both**

Where appropriate, preferred choices are in **bold** font

Table 7.13 (*Continued*)

Constraint	Proper constraint verified?	**Yes*** No
If over-constrained	In opposition?	**Yes** No
	Redundant features?	Yes **No**
Compatibility	Between basic shapes and assembly motion?	**Yes** No
	Between locator pairs and assembly motion?	**Yes** No
	Between lock pairs and assembly motion?	**Yes** No
	Assembly and disassembly motions are the same?	**Yes** No
Required enhancements	Guides?	**Yes** No
	Operator feedback?	**Yes** No
	Clearance?	**Yes** No
	Process friendly?	**Yes** No
Desirable enhancements	Compliance?	**Yes** No
	Fine-tuning?	**Yes** No
Other enhancements (depends on application)	Pilots	**Yes** No
	Guards	Yes No
	User feel?	Yes No
	Visuals?	Yes No
	Assists?	Yes No
	Retainers?	Yes No
	Back-up locks?	Yes No
Operator feedback	Preferred	**Tactile** Audible Visual **Multiple**
Feature design (depends on application)	Based on analysis?	**Yes** No
	Rules of thumb?	**Yes** No
Evaluation of first parts	Assembly interference?	**Yes** No
	Acceptable assembly force?	**Yes** No
	Feature damage during assembly?	**Yes** No
	Operator feedback?	**Yes** No
	Compatibility and constraint?	**Yes** No
	Attachment durability?	**Yes** No
	User feel?	**Yes** No
	Part and feature consistency?	**Yes** No
Fine-tuning	Location?	**At**/opposite critical alignment sites.
Compliance	Location?	**Within**/between constraint pairs. At/**opposite** critical alignment sites. At/**opposite** critical load bearing sites.
Parts	Sharp corners?	Yes **No**
	Thick sections?	Yes **No**
	Sudden section changes?	Yes **No**

Where appropriate, preferred choices are in **bold** font.

- By using basic shapes and combinations of assembly motions and engage directions, the suggested development process encourages and enables improved spatial reasoning and creativity by designers when developing snap-fit attachment concepts.
- With but minor changes, the attachment level development process for snap-fits is applicable to other attachment development and design situations. Simply treat the lock selection step as allowing any mechanical fastener as a locking option rather than limiting the selection to just integral lock features.
- The process will encourage the designer to generate truly different concepts, not just variations on one theme. Take advantage of this opportunity. Having variations on one attachment theme can be useful, but focusing only on one theme will limit creativity.
- If a snap-fit is indicated, play it safe and over-design the attachment if necessary. The design will still save money over conventional fasteners.
- The designers must get their hands and their spatial reasoning involved in the creative process by making sketches and by building models.
- Get others involved in the development process; many people think snap-fits are fun, especially if they have parts to play with.

References

1. Wilson, Frank R., 1998, *The Hand*, p. 291, Pantheon Books, New York.
2. Boothroyd, G., *Design for Manufacture and Life-Cycle Costs*, 1996, SAE Design for Manufacturability TOPTEC Conference, Nashville, TN.
3. Porter, C.A., Knight, W.A., *DFA for Assembly Quality Prediction during Early Product Design*, (1994), Proceedings of the 1994 International Forum on Design for Manufacture and Assembly, Newport, RI. Boothroyd Dewhurst, Inc., Wakefield, RI.
4. Malloy, Robert A. 1994, *Plastic Part Design for Injection Molding*, Hanser/Gardner Publications, Inc., Cincinnati OH.
5. Ford, R.B., Barkan, P., 1995, *Beyond Parameter Design—A Methodology Addressing Product Robustness at the Concept Formation Stage*, 1995 National Design Engineering Conference, Chicago, IL.
6. Meeker, D.G., 1994, *Benchmarking, Its Role in Product Development*, Proceedings of the 1994 International Forum on Design for Manufacture and Assembly, Newport, RI, Boothroyd Dewhurst, Inc., Wakefield, RI.
7. Michalko, M., 1998, *Thinking Like a Genius*, THE FUTURIST, May 1998, pp. 21–25.
8. Woodsen, W.E., 1981, *Human Factors Design Handbook*, New York: McGraw-Hill Book Company.
9. Marras, W.S., 1997, *Biomechanics of the Human Body*, pp. 233–265, Handbook of Human Factors and Ergonomics, G. Salvendy, Ed. New York: John Wiley & Sons, Inc.
10. Karwowski, W., Marras, W. S., 1997, *Work-related Musculoskeletal Disorders of the Upper Extremities*, pp. 1124–1172, Handbook of Human Factors and Ergonomics, G. Salvendy, Ed. New York: John Wiley & Sons, Inc.
11. Tres, P., 2000, *Designing Plastic Parts for Assembly* Hanser/Gardner Publications, Inc., Cincinnati OH.
12. *Designing with Plastic—The Fundamentals, Design Manual TDM-1*. 1996, Ticona LLC (formerly Hoechst Celanese), Summit, NJ.
13. Bonenberger, Paul R., 1998, *An Attachment Level Design Process for Snap-Fit Applications*, DE-Vol. 99, MED-Vol. 7, 1998 ASME International Mechanical Engineering Congress and Exposition, Anaheim, CA.

8 Diagnosing Snap-Fit Problems

The goal of careful design is always to prevent problems from occurring in the first place but snap-fits do sometimes fail. They also have their share of assembly and usage problems. A full understanding of the various failure modes and their relationship to the most likely root causes can help one more quickly diagnose and solve problems. This, of course, minimizes the cost and time impact of fixing a problem but it also helps ensure that the proposed changes will indeed fix the problem. Nothing is worse than making changes to a product and finding that the problem still exists or has even gotten worse. Accurate diagnosis is particularly valuable during product development when prototype testing may indicate the need for improvements, yet time and cost constraints limit the available options.

8.1 Introduction

The root causes of many snap-fit problems are at the attachment level. Yet, many times, the first attempts to fix the problems are at the feature level, thus are doomed to failure or to cost much more than they should. When evaluating any snap-fit problem, *even feature failure or damage*, first verify that all attachment level requirements have been satisfied. If not, address them before attempting a feature level fix [1].

First, it is important to define "problem" because the term includes much more than simply feature breakage. A snap-fit problem is identified by the following symptoms:

- Difficult assembly
- Short-term feature failure or damage
- Long-term feature failure or damage
- Part distortion or damage
- Part loosening and/or squeaks or rattles
- Unintended part release
- Service difficulty
- Customer complaints about ease of operation

It is important to remember that these are symptoms of a problem; they are not the root cause of the problem. Simply treating the symptom may not fix the real problem or it may create other problems in the attachment. Many of the above symptoms can have both attachment and feature level root causes. Sometimes, the root cause turns out to be a combination of several shortcomings.

Figure 8.1 reflects the author's personal experience in trouble-shooting snap-fit problems in products. Note the high incidence of causes related to attachment level issues and the high frequency of multiple root causes for problems.

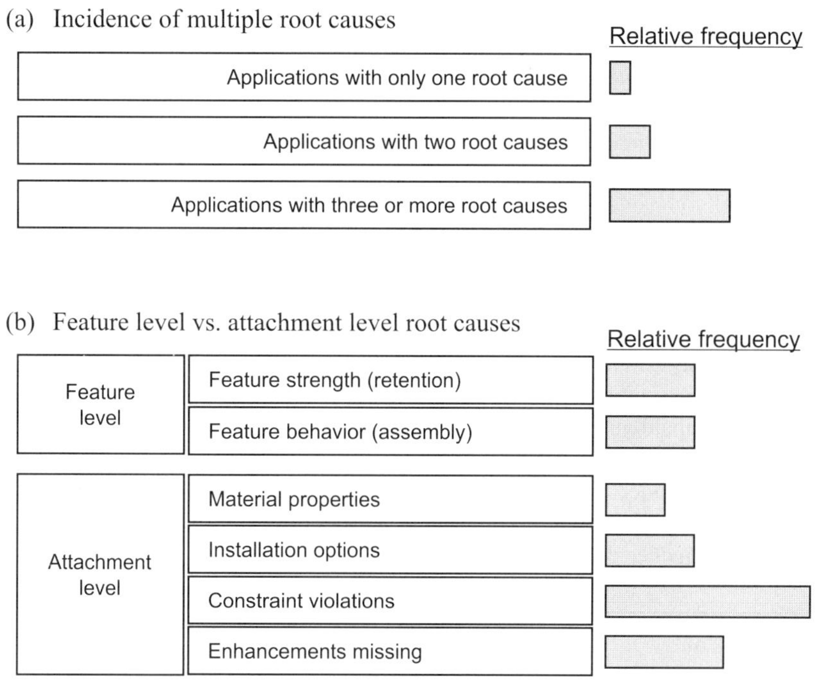

(a) Incidence of multiple root causes

Relative frequency

| Applications with only one root cause |
| Applications with two root causes |
| Applications with three or more root causes |

(b) Feature level vs. attachment level root causes

Relative frequency

Feature level	Feature strength (retention)
	Feature behavior (assembly)
Attachment level	Material properties
	Installation options
	Constraint violations
	Enhancements missing

Figure 8.1 General trends in snap-fit problems

8.1.1 Rules for Diagnosing Snap-Fit Problems

- Do not mistake a symptom for a root cause. This is important in any problem-solving effort.
- Always remember that the root cause of many, if not most, snap-fit problems is at the attachment level, not the feature level. Corollary to this rule is to also remember that many attachment level problems are related to improper constraint.
- Resolve all attachment level causes of a problem before attempting any feature level fixes. Recall that some problems are a combination of both feature and attachment level causes.
- Always try the easiest fixes first.
- Be aware that most feature level changes in a snap-fit will have multiple effects. A change to fix one problem is very likely to change other behaviors and may create new problems.

8.1.2 Mistakes in the Development Process

Tracing back through the development process can sometimes give clues as to the root cause of a problem. Most problems, whether they are attachment or feature level, are the result of mistakes made during the development process. Some of the more common mistakes are:

- Improper constraint in the attachment, characterized by:
    Over-constraint where features "fighting" each other can cause breakage during assembly or in service due to thermal expansion. Over-stress due to residual assembly forces can cause long-term failure.
- Under-constraint where: Features are carrying the wrong loads or excessive loads. Weak or compliant parts are expected to provide a rigid base for lock or locator features.
- Failure to fully consider material properties, including:
    Incomplete material property data available.
    Failure to consider plastic creep and thermal effects.
- Failure to anticipate assembly variables such as:
    Incompatibilities involving engage direction, assembly motion, feature positioning and feature style.
    Designed-in assembly frustration and difficult assembly.
- Failure to anticipate all possible end-use conditions, including:
    Failure to consider disassembly and service.
    Failure to consider all loads including unexpected or improper but possible load conditions, (such as dropping or striking a product).
    Failure to consider customer usage.

8.2 Attachment Level Diagnosis

Attachment level problems are often independent of the lock. In other words, they would be occurring regardless of the locking features style used. Understanding the key requirements of constraint, compatibility and robustness can help one recognize and resolve many attachment level problems. Remember too that certain enhancements are required for every application. Always verify the presence of these four enhancements, if any are missing, problems are likely. They are:

- Guides
- Clearance
- Operator feedback
- Process-friendly

The four most common symptoms related to attachment level problems are:

- Difficult assembly
- Parts distorted
- Feature damage
- Loose parts

For each of these symptoms, the most likely root causes are listed below.

8.2.1 Most Likely Causes of Difficult Assembly

- Over-constraint
- Assembly motion and constraint feature incompatibility
- Basic shape and assembly motion incompatibility
- Access and basic shape incompatibility
- Access and assembly motion incompatibility
- Parts warped
- Simultaneous engagement of several features
- No guide or clearance enhancements
- No operator feedback and/or feedback interference
- Mating part is hard to hold or handle

8.2.2 Most Likely Causes of Distorted Parts

- Parts warped when made
- Distorted in assembly
- Feature tolerances and position robustness
- Over-constraint
- Compliant (flexible) parts, often panels are not constrained at enough points

8.2.3 Most Likely Causes of Feature Damage

Feature damage does not necessarily indicate a feature problem. This is one of the most common errors in diagnosis. Many times feature damage is a symptom, not the root cause.

- Over-constraint
- Under-constraint
- Incompatibility between features and assembly motion
- Long-term creep or yield
- Damaged during assembly (see Difficult Assembly)
- Damaged during shipping and handling
- Poor processing, not process-friendly
- Abuse in usage
- Abuse or damage during service/removal
- Missing guide or clearance enhancements

8.2.4 Most Likely Causes of Loose Parts

- Feature damage (see above)
- Weak feature mounting area(s) on mating and base parts
- Difficult assembly (see above)
- Under-constraint
- Compliant parts do not provide a strong base for the constraint features

8.3 Feature Level Diagnosis

Only after all attachment level root causes are either fixed or ruled out can we begin to consider feature level root causes for the problem. Sometimes, new parts must be produced that reflect all the attachment level fixes before feature level causes can be identified. Recall the panel-to-opening-application example in Section 4.

Obviously, however, it is desirable to identify and make feature level changes before new parts are made. If the problem is indeed a feature problem, simple changes to the feature dimensions may be possible. These are the easiest changes to make. If they do not fix the problem, then more difficult changes to the lock feature style or to the lock pair are indicated. The most common feature level problems are:

- High assembly force, see Table 8.1 for recommended fixes.
- High feature strain or feature damage during assembly or disassembly, see Table 8.2.
- Low retention strength, and lock damage under loads, see Table 8.3.
- High separation force, see Table 8.4.

In Tables 8.1 through 8.4, the fixes are listed from top to bottom beginning with those that are easier to implement and moving down through the changes that are more difficult or costly. Ease of implementation was based on the following reasonings:

- Changes to the lock retention mechanism are generally the easiest.
- Changes to the lock deflection mechanism are generally more difficult.
- Changes to the attachment system are generally the most difficult.

The recommended changes and the predicted interactions in these tables are written primarily with the cantilever beam style lock in mind. However, many of the changes apply to all lock styles.

We know that fixing one feature problem may create another. For example, making a cantilever hook lock stronger to solve a problem with low retention strength may increase the assembly force and may also increase the strain in the hook. If the assembly force becomes too high or strain is excessive, a new set of problems will surface. Whenever a change is proposed, it is important to understand these interactions to avoid creating other problems.

8.3 Feature Level Diagnosis 259

Table 8.1 Feature Level Solutions for High Assembly Force

Ease	Make change to	Recommended change	Feature strain or damage during assembly or disassembly	Retention strength or lock damage under loads	Separation force	Interactions +	Interactions −
1	Retention mech.	Reduce insertion face angle	—	—	—	1	0
1	Retention mech.	Add contour to insertion face	—	—	—	1	0
1	Retention mech.	Add dwell surface to catch	—	—	—	1	0
1	Retention mech.	Make retention face shallower (decrease deflection)	reduce	worse	reduce	3	1
2	Deflection mech.	Make beam longer	reduce	worse	reduce	3	1
2	Deflection mech.	Reduce beam thickness overall	reduce	worse	reduce	3	1
2	Deflection mech.	Reduce beam thickness at end by tapering	reduce	worse	reduce	3	1
2	Deflection mech.	Reduce beam width overall	—	worse	reduce	2	1
2	Deflection mech.	Reduce beam width at end by tapering	—	worse	reduce	2	1
3	Locking system	Decouple insertion and retention behaviors	reduce	improved	reduce	4	0
3	Locking system	Design for sequential lock engagement	—	—	—	1	0
3	Locking system	Redesign for a tip assembly motion	—	—	—	1	0
3	Locking system	Decrease mating feature stiffness (increase deflection)	reduce	worse	reduce	3	1
3	Locking system	Make base area more flexible (Q-factor)	reduce	worse	reduce	3	1
3	Locking system	Change lock style	—	—	—	1	?
3	Locking system	Change part material	—	—	—	1	?

A "—" in the effects column indicates either no effect or effect cannot be predicted.

Table 8.2 Feature Level Solutions for High Feature Strain or Damage During Assembly or Disassembly

Ease	Make change to	Recommended change	Reducing *high feature strain* may also have these effects:			Interactions		
			Assembly force	Retention strength or lock damage under loads	Separation force	+	−	?
1	Process	Verify part manufacturing process is correct	—	—	—	?	—	?
1	Retention mech.	Make retention face shallower (decrease deflection)	reduce	worse	reduce	3	1	
2	Deflection mech.	Make beam longer	reduce	worse	reduce	3	1	
2	Deflection mech.	Reduce beam thickness overall	reduce	worse	reduce	3	1	
2	Deflection mech.	Reduce beam thickness at end by tapering	reduce	worse	reduce	3	1	
2	Deflection mech.	Increase beam thickness at base by tapering	increase	improved	increase	2	2	
3	Locking system	Verify part design is process-friendly	—	—	—	?	?	
3	Locking system	Decouple insertion and retention behaviors	reduce	improved	reduce	4	0	
3	Locking system	Add guidance enhancement feature	—	—	—	1	0	
3	Locking system	Add visual enhancement feature	—	—	—	1	0	
3	Locking system	Decrease mating feature stiffness	reduce	worse	reduce	3	1	
3	Locking system	Make base area more flexible (Q-factor)	reduce	worse	reduce	3	1	
3	Locking system	Add guard enhancement feature	increase	—	increase	1	2	
3	Locking system	Change lock style	—	—	—	1	?	
3	Locking system	Change part material	—	—	—	1	?	

A "—" in the effects column indicates either no effect or effect can not be predicted.

Table 8.3 Feature Level Solutions for Low Retention Strength or Lock Damage Under Load

Ease	Make change to	Recommended change	Changes to fix *low retention strength or lock damage under load* may also have these effects:			Interactions	
			Feature strain or damage during assembly or disassembly	Assembly force	Separation force	+	−
1	Retention mech.	Load beam closer to neutral axis	—	—	—	1	0
1	Retention mech.	Increase retention face angle	—	—	increase	1	1
1	Retention mech.	Add contour to the retention face	—	—	increase	1	1
1	Retention mech.	Make retention face deeper (increase deflection)	increase	increase	increase	1	3
2	Deflection mech.	Increase beam thickness at base by tapering	reduce	increase	increase	2	2
2	Deflection mech.	Increase beam width at base by tapering	—	increase	increase	1	2
2	Deflection mech.	Make beam shorter	increase	increase	increase	1	3
2	Deflection mech.	Increase beam thickness overall	increase	increase	increase	1	3
2	Deflection mech.	Increase beam width overall	increase	increase	increase	1	3
3	Locking system	Decouple insertion and retention behavior	reduce	reduce	reduce	4	0
3	Locking system	Reorient lock to carry less load	—	—	—	1	0
3	Locking system	Add more lock features	—	increase	increase	1	2
3	Locking system	Add retainer enhancement feature	—	increase	increase	1	2
3	Locking system	Increase mating feature stiffness	increase	increase	increase	1	3
3	Locking system	Make base area less flexible (Q-factor)	increase	increase	increase	1	3
3	Locking system	Change lock style	—	—	—	1	?
3	Locking system	Change material	—	—	—	1	?

A "—" in the effects column indicates either no effect or effect can not be predicted.

Table 8.4 Feature Level Solutions for High Separation Force

Ease	Make change to	Recommended change	Feature strain or damage during assembly or disassembly	Assembly force	Retention strength or lock damage under loads	Interactions +	Interactions −
1	Retention mech.	Make retention face shallower (decrease deflection)	reduce	reduce	worse	3	1
1	Retention mech.	Reduce retention face angle	—	—	worse	1	1
2	Deflection mech.	Make beam longer	reduce	reduce	worse	3	1
2	Deflection mech.	Reduce beam thickness overall	reduce	reduce	worse	3	1
2	Deflection mech.	Reduce beam thickness at end by tapering	reduce	reduce	worse	3	1
2	Deflection mech.	Reduce beam width overall	—	reduce	worse	2	1
2	Deflection mech.	Reduce beam width at end by tapering	—	reduce	worse	2	1
3	Locking system	Decouple insertion and retention	reduce	reduce	improved	4	0
3	Locking system	Add assist enhancement feature	—	—	—	1	0
3	Locking system	Decrease mating feature stiffness (increase deflection)	reduce	reduce	worse	3	1
3	Locking system	Make base area more flexible (Q-factor)	reduce	reduce	worse	3	1
3	Locking system	Change lock style	—	—	—	?	?
3	Locking system	Change material	—	—	—	?	?

A "—" in the effects column indicates either no effect or effect can not be predicted.

Within each ease-of-change group in the tables, the suggested changes are ranked by the number of additional positive or negative interaction they may have on the attachment. Usually a negative effect is simply an incremental shift in a particular characteristic. A negative effect does not guarantee a new problem, just a movement toward a condition that will increase the likelihood of a problem. For each of the four types of problems addressed by the tables, these interactions were developed in terms of the other three problems.

8.4 Summary

This chapter described an attachment level approach to diagnosing and fixing the most common snap-fit problems. Problems were first defined as a broad range of situations including, but not limited to, the more obvious ones involving snap-fit feature damage and failure. Most importantly, an approach of addressing systemic causes before attempting feature level fixes is explained. The snap-fit diagnostic process is summarized in Fig. 8.2.

8.4.1 Important Points in Chapter 8

- Do not mistake a symptom for a root cause.
- The root cause of many, if not most, snap-fit problems is at the attachment level, not the feature level.

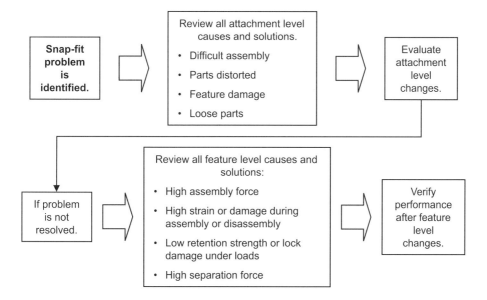

Figure 8.2 The diagnostic process for snap-fits

- Do not assume that a feature failure has a feature level root cause.
- Many attachment level problems result from improper constraint.
- Resolve all attachment level causes of a problem before attempting any feature level fixes.
- Some problems will be combination of both feature and attachment level causes.
- Always try the easiest fixes first.
- Most feature level changes in a snap-fit will have multiple effects. A change to fix one problem is very likely to change other behaviors and may create new problems.

Reference

1. Bonenberger, P.R., (1999), *Solving Common Problems in Snap-Fit Designs*, Western Plastics Expo, Jan, 1999, Long Beach, CA.

Index

Action (function), 27
Activation, 44, 109
Activation enhancements, 44, 109
Adjustable inserts, 125, 126
Adjustments to calculations, 187–197
Alignment requirements, 62, 127, 145
Alternative fasteners, 37, 246
Analysis, 69, 94, 162, 198, 216, 243
Analysis example, 201–208
Annular (lock), 64, 68, 91
Application appropriate for snap-fit, 224–228
Assembly
   –enhancements, 44, 96, 109
   –force, 29, 106, 131, 162, 206
   –force signature, 79, 85, 106
   –motion, 38–40, 61, 62, 235, 237, 257
Assists (enhancement), 44, 109, 111, 113, 133
Assumptions for analysis, 164, 198
Attachment Level
   –and design for assembly, 10
   –and other attachments, 2, 11
   –Construct, 1, 14, 223
   –definition of, 6, 7, 47
   –history, viii
   –problem symptoms, 256
   –vs. feature level, 6
Attachment type, 27

Back-up lock (enhancement), 44, 114, 119, 133
Base part, 31
Basic shapes
   –and assembly motion, 40
   –defined, 29, 31, 257
   –frequency in applications, 229
   –tables, 34
Beam, 68, 155, 178, 199, 209
   –terminology, 179
Benchmarking, 95, 230
   –checklist, 231
   –rules, 231
Bezel applications, 157–159
Bi-directional forces, 145

Cantilever hook, rules of thumb, 178
Cantilever lock, 68, 84
Catch (locator), 41, 50
Cavity (basic shape), 32, 34
Checklists (worksheets)
   –application appropriate for snap-fit, 225
   –benchmarking, 231

   –best concept, 244
   –constraint worksheet examples, 143–149
   –constraint worksheet original, 240
   –feature problem diagnosis, 259–262
   –final evaluation, 251, 252
Clearance (enhancement), 46, 98, 131, 256
Coefficient of friction, 172, 174
Coefficient of Linear Thermal Expansion, 176, 177
Compatibility, 20, 22
Compliance (enhancement), 44, 67, 115–119, 133, 146
Concept development, 219
Cone (locator), 49
Constant section beam, 199–209
Constraint, 17, 40, 47, 57, 92, 135–150, 160, 337, 240
   –features, 14, 47, 92
   –improper, 19, 46, 139, 255
   –pairs, 19, 145, 237
   –proper, 20, 139
   –rules, 141, 239
Cost, 101, 121
Couples, 63, 66
Creativity, 4, 218, 232
Creep, 175
Crush ribs (enhancement), 117, 118
Cutout (locator), 52

Darts (enhancement), 117, 118
Decoupling, 135, 151–160
   –levels of, 153
   –summary, 159
Deflection
   –force, 204, 208
   –magnification factor, 189–194
   –magnification factor tables, 192, 193
   –of mating feature/part, 188, 191–194, 205
Deflection-thickness ratio, 181
Degrees of Motion (DOM), 17, 58, 136, 139
   –and locator pairs, 57
   –maximize removal of, 60, 62, 142
Descriptive elements, 25
Design for assembly
   –process, 10
   –practitioners, 10
Design point, 168–171
Design rules, 46, 55, 93, 120, 131, 141, 145, 160, 178, 231, 233, 236, 237, 239, 246, 250, 255
Diagnosis, 135, 254,
   –at the attachment level, 256
   –at the feature level, 258
   –tables, 259–262

Difficult assembly, causes of, 257
Dimensional
 –control, 62
 –robustness, 116
Distorted parts, causes of, 257
Draft angle, 123
DTUL, 175

Edge (locator), 51
Effective angle, 188, 194, 206
Efficiency, lock 89
Elastic limit, 167
Elasticity (enhancement), 119
Elements, 13, 14, 25, 26, 43, 44, 45
 –and the development process, 224
Enclosure (basic shape), 32, 34
Engage direction, 35, 145, 233, 236
Enhancements, 7, 11, 14, 43, 95–134, 233, 242
 –and the development process, 132
 –for activation, 43, 109
 –for assembly, 43, 96, 108
 –for manufacturing, 43, 120
 –for performance, 43, 114
 –required, 128, 131, 256
 –summary table, 45, 129
Ergonomics, 37, 131, 237
Evaluation
 –initial strain, 185
 –final, 251–252
Examples, 30, 100–104, 108, 150

Fasteners, 227, 228, 245–249
Feature
 –analysis, 163
 –damage, causes of, 109, 115, 257
Feature level, 11, 12
 –diagnosis, 258–263
 –vs. attachment level, 6
Feedback (enhancement), 104, 131, 256
Final (attachment type), 28
Fine-tuning, 44, 64, 121, 125, 127, 133, 147, 250
Finite element analysis, 198
Fixed snap-fits (action), 27
Forces, 36, 81, 235
Function, 27–30, 91, 229
 –summary, 30

Gates, 73, 124
Guards (enhancement), 44, 112, 114
Guidance (enhancement), 43, 96, 109
Guides (enhancement), 97, 103, 131, 256

Hands on, 8, 237
High assembly force, 80, 109, 258

High assembly force, fixes for, 259
High separation force, fixes for, 262
High strain or damage in features, fixes for, 260
Hole (locator), 52
Hook (lock), 178, 199, 209, 211
Hook styles, 68, 75, 76

IBM, 8
Impact force or load, 28, 36, 168, 198
Improper constraint, 19, 46, 135, 139, 255
Insertion face, 78, 80
 –angle, 87, 153, 181, 195, 206
 –profile, 78, 80, 105, 113, 212
Insertion force signature, 79, 80, 85, 106, 214
Inserts, adjustable, 125
Integral attachment, vii, 4
Isolators (enhancement), 119

Key requirements, 14, 16, 25, 45
Knitlines, 73

Land (locator), 41, 50, 64
Length of beam, 178
Length to thickness ratio, 180, 191
Level 0, (decoupling), 153
Level 1, (decoupling), 154
Level 2, (decoupling), 155
Level 3, (decoupling), 156
Level 4, (decoupling), 156
Line-of-action, 58, 59, 65
Living hinge (locator), 54
Local yield (enhancement), 117
Locators, 40, 41, 47, 48, 55, 67, 74, 92, 93
 –design rules, 55
 –pairs, 57, 61, 63
 –styles, 48
 –summary, 56
Lock type (function), 29
Locks, 42, 43, 44, 67, 68, 74, 84, 85, 119, 139
 –alternatives, 246
 –damage, fixes for, 262
 –decoupling, 151, 159, 160
 –efficiency, 89, 90, 92, 159
 –pairs, 42, 77, 91
 –styles, 67, 68, 75, 76
Loops (lock), 70, 72, 74, 93, 157
Loose
 –fasteners, 120, 245
 –parts, causes of, 258
Low retention strength, fixes for, 261
Lug
 –as a locator, 41, 48
 –as a lock, 74, 77, 78

Manual or non-releasing locks (lock type), 30, 72, 88, 115
Manufacturing enhancements, 44, 120, 121
Material properties, 163
 –other effects, 175
 –sources for, 164
Mating feature deflection, 188, 191–194, 205
Mating part, 31
 –deflection, 188, 191–194, 205
Maximize DOM removed, 58, 60, 142
Maximum
 –allowable strain, 173, 201, 216
 –assembly force, 206, 208
Mechanical advantage, 64
Metal clips, 245, 249
Metal-safe (enhancement), 125
Metal-to-metal snap-fits, 5
Metal-to-plastic snap-fits, 5
Models, use of, 238
Mold adjustment, 44, 125
Moveable (action), 27
Multiple concepts, 232

Natural locator, 40, 41, 48
Nesting, 40, 161
Noise
 –background, 105, 107
 –squeak and rattle, 116
Non-permanent (retention), 29
Non-releasing or manual locks (lock type), 30, 72, 88, 115

Opening (basic shape), 32, 34
Opposing features, 140
Other attachments, 2, 11
Over-constraint, 20, 142, 160, 161, 241, 256, 257
 –opposing features, 140
 –redundant features, 140

Panel (basic shape), 31, 34
Perfect constraint, 136, 139, 148, 150, 160
Performance enhancements, 44, 114
Permanent locks (retention), 28
Physical elements, 25
Pilot (enhancement), 99
Pin (locator), 50
Pivot (assembly motion), 38, 142
Planar (lock), 68, 84
Plastic push-in fasteners, 245, 248
Polaroid, 8, 134
Process-friendly (enhancement), 44, 121, 133, 256
Proper constraint, 20, 25, 46, 139, 160, 163
Proportional limit, 166
Prototypes, 238

Protrusion spacing, 123
Push (assembly motion), 38, 60, 142, 236
Push-in fasteners, 245, 248

Q-factor, 191

Radius, 98, 122, 124
Redundant features, 140
Release behavior, 71, 207
Releasing lock (lock type), 73
Required enhancements, 128, 131, 256
Retainers (enhancement), 84, 115, 159
Retention, 28, 67, 68, 70, 80, 82, 84, 87, 90, 156, 157, 162, 181, 182, 183, 196, 207, 208, 243, 261
Retention (function), 28
Retention
 –mechanism, 70
 –signature, 83, 85
 –strength, 71, 72, 80, 115, 157, 162, 207
Retention face, 80, 182, 215
 –angle, 80, 182, 196
 –depth, 181
 –profile, 80, 83, 215
Robustness, 21, 23, 25, 43, 116
Rules of thumb, cantilever hook, 178

Sample parts, 8
Screws, 245, 248
Secant modulus, 167, 172
Section properties, 199
Section changes, 123, 124
Separation, 71, 81, 145, 181, 207
 –direction, 35
 –force, 60, 81, 145, 207, 262
Separation force, fixes for high, 262
Shut-off angle, 123, 124
Side-action hook, 109, 155, 160
Sinkmarks, 124
Slide (assembly motion), 38, 142
Slot (locator), 52
Snap-fit, 3, 4
 –attachment level definition of, 6, 7, 47
 –feature level definition of, 2
 –structural, 5
 –vs. threaded fastener, 5, 224, 228
Solid (basic shape), 31, 34
Sources of materials data, 163
Spatial
 –elements, 25
 –reasoning, 4
Spring clips, 250
Strain, 169, 172, 173, 199, 201, 202, 203, 208, 260
Strain limit, 185
Strength, 16, 17, 57, 128, 133, 169, 261

Stress concentration, 188, 202
Stress-strain curve, 164, 165, 175
Structural snap-fits, 5
Surface
–as a basic shape, 32
–as a locator, 34, 50
Symbols, as visual enhancements, 111

Tab (locator), 48
Tapered
–beams, 209
–features, as compliance enhancements, 117
Temporary lock (attachment type), 28
Terminology, 55, 179, 200
Thermal effects, 141, 142, 176, 177
Thick sections, 123, 124, 175
Thickness tapered beam, 209
Thickness and width tapered beam, 212
Threaded fasteners, 245–248
–vs. snap-fit, 5, 228
Threshold angle, 183
Tip (assembly motion), 38, 60, 98, 142, 236
Tolerance, 21, 44, 63, 114, 116, 125, 128, 133, 136, 139, 163, 177, 257
Torsional lock, 64, 68, 90
Track (locator), 48, 49
Trap (lock), 64, 68, 85–87, 89, 157

Twist (assembly motion), 38, 142

Ultimate strength, 167
Under-constraint, 20, 139, 241, 256, 257, 258
Unintended release (or separation), 29, 36, 81, 182, 207, 254
User-feel (enhancement), 113
Visuals (enhancement), 44, 109, 110

Wall
–deflection, 189–191
–thickness, 122
Wedge (locator), 49
Width of beam, 184
Worksheets (checklists)
–application appropriate for snap-fit, 225–227
–benchmarking, 231
–best concept, 244
–constraint worksheet examples, 143–149
–constraint worksheet original, 240
–feature problem diagnosis, 258–263
–final evaluation, 251, 252
Width tapered beam, 211

Yield
–point, 167
–strength, 167

Paul Bonenberger is a Staff Project Engineer with the General Motors North American Engineering Center. He has a background in final assembly and product test and development and has spent considerable time as a subject matter expert in the field of mechanical attachments. He created the Attachment Level™ Construct that is the basis for this book. Having taught threaded fastener and snap-fit classes at General Motors for many years, he now participates in the University of Wisconsin CCEE course "Snap-Fits and Product Design" and other technical training events.